Friedrichshafen Aircraft of WWI

A Centennial Perspective on Great War Airplanes

Jack Herris

Great War Aviation Centennial Series #21

This book is dedicated to friend and colleague Jim Miller for his contributions as an illustrator supplying excellent profiles for many Aeronaut Books.

Acknowledgements

My sincere thanks to Colin Owers, Bill Toohey, Michael Schmeelke, Greg VanWyngarden, Dave Hooper, Dave Watts, Richard Andrews, and F. Gerdessen for photographs, Bob Pearson for his color profiles, Aaron Weaver for photographs and image editing, Marty Digmayer for scale drawings, and the Museum of Flight in Seattle for photographs. Thanks also to Steve Anderson for the original cover painting specifically for this book and Aaron Weaver for the cover design. I also want to thank Reinhard Zankl for his helpful information regarding production orders and serial numbers and for photographs. Any errors are my responsibility.

Color aircraft profiles © Bob Pearson. Purchase his CD of WWI aircraft profiles for $50 US/Canadian, 40 €, or £30, airmail postage included, via Paypal to Bob at: **bpearson@kaien.net**

Cover painting by Steve Anderson. Please see Steve's website at: **www.anderson-art.com**

For our aviation books in print and electronic format, please see our website at: **www.aeronautbooks.com**. I am looking for photographs of the less well-known German aircraft of WWI. For questions or to help with photographs you may contact me at **jherris@me.com**.

ISBN: **978-1-935881-35-3**

Design and layout: Jack Herris
Cover design: Aaron Weaver
Digital photo editing: Aaron Weaver & Jack Herris

Aeronaut Books

www.aeronautbooks.com

Table of Contents

Forward

The purpose of this book is to document Friedrichshafen aircraft of WWI. Friedrichshafen was the largest manufacturer of floatplanes for the German Navy and also produced a successful series of G-type, twin-engine bombers that eventually outnumbered the bombers from other manufacturers at the front. In addition to seaplanes and twin-engine bombers, Friedrichshafen produced prototypes of land-based fighters and reconnaissance airplanes, none of which reached production.

Friedrichshafen used internal factory designations for its designs, but not all designs were built, leaving gaps in the numbering system. Naval aircraft types were known by their factory designations but Army types were given official military type designations. Where known, both designations are given. The *SVK* ordering and acceptance tables for Friedrichshafen are reproduced here as given in the original documents but are redrawn for legibility and known errors and omissions have been corrected.

Friedrichshafen

Flugzeugbau Friedrichshafen GmbH was founded July 17, 1912 by Theodor Kober, previously an employee of the Zeppelin company. Count Zeppelin was one of two shareholders, and Kober was the sole Director. The company was originally located in and around the old floating Zeppelin Hall in Manzell near Friedrichshafen, the same town where Zeppelins were built, on Lake Constance (the *Bodensee* in German) on the Rhine. Lake Constance is bordered by Bavaria and Baden-Württemberg in Germany plus Austria and Switzerland. The company later expanded beyond these facilities as it grew during the war.

Friedrichshafen became one of the most important WWI aircraft manufacturers in Germany, growing from 149 employees in 1914 to 1,508 by the end of 1917 and 3,240 by the Armistice.

Friedrichshafen's location on Lake Constance was convenient for its main business of designing and building floatplanes for the German Navy. Friedrichshafen's floatplanes earned a good reputation for robustness and reliability, essential assets for aircraft flown over water, and were able to operate in higher sea states than most other German floatplanes. These qualities were critical to naval fliers operating in the rough northern seas around Germany and led to Friedrichshafen's success as the largest supplier of seaplanes to the German Navy.[1]

In addition to floatplanes, Friedrichshafen also produced a successful series of twin-engine bombers for the Army. Friedrichshafen's bombers, like their seaplanes, were robust aircraft with good handling qualities. In particular, the Friedrichshafen bombers did not suffer the aft center-of-gravity problems that afflicted some other bomber types and were therefore were much safer to land at night.

Although Friedrichshafen produced several single-engine landplane designs for the Army, none were selected for production.

[1] Through the end of December 1918, the German Navy received about 2,365 seaplanes. Of that number, Friedrichshafen delivered 854 (36%). Hansa-Brandenburg delivered 671 seaplanes (28%), Albatros delivered 165 (7%), Sablatnig delivered 160, Gotha delivered 145, and Rumpler delivered 129. *Flugzeugbau Friedrichshafen GmbH* attributes 42% of naval aircraft (908 of 2150) to Friedrichshafen.

Above: Friedrichshafen's first airplane design, the primitive FF1. Appropriately, it was a seaplane as were most Friedrichshafen designs. The FF1 first flew in mid-November 1912, the year that the company was founded.

Above: Theodor Kober in 1894; he founded Friedrichshafen in 1912.

Above: The primitive Friedrichshafen's FF1 in flight. (The Peter M. Bowers Collection/The Museum of Flight)

Above: Friedrichshafen suffered a disaster on the night of 13/14 April 1918 when a major fire destroyed a large part of the Manzell factory including the old Zeppelin Hall, where seven aircraft were also destroyed. Numerous design drawings. were also burned, delaying development of future prototypes. Coincidentally, all the fire hydrants were unusable, and sabotage was strongly suspected but not proven; all evidence had been burned. This was the result.

Friedrichshafen Aircraft Actually Built		
Company Designation	Type (Biplane unless Noted)	Role & Status
FF1	Pre-war pusher floatplane	Prototype only
FF2	Pre-war tractor floatplane	Prototype only
FF4	Pre-war tractor floatplane	Prototype only
FF6	Pre-war tractor biplane	Prototype only
FF8	Pre-war tractor floatplane	Prototype only
FF9	Pre-war pusher floatplane	Prototype only
FF11	Pre-war pusher flying boat	Prototype only
FF15	Pre-war pusher floatplane	Prototype only
FF17	Pre-war tractor floatplane	Prototype only
FF19	Pre-war tractor floatplane	Reconnaissance, production
FF21	Pre-war pusher flying boat	Prototype only
FF27	Pre-war pusher floatplane	Prototype only
FF29	Tractor floatplane	Reconnaissance, production
FF31	Pusher floatplane	Reconnaissance prototype
FF33 Series	Tractor floatplane	Reconnaissance, production
FF34	Pusher floatplane	Reconnaissance prototype
FF35	Twin-engine pusher floatplane	Torpedo bomber prototype
FF36 (Army Desig.: G.I)	Twin-engine pusher bomber	Bomber, prototype
FF37	Pusher landplane	Reconnaissance prototype
FF38 (Army Desig.: G.II)	Twin-engine pusher bomber	Bomber, production
FF39	Tractor floatplane	Reconnaissance, production
FF40	Tractor floatplane	Reconnaissance prototype
FF41A & FF41AT	Twin-engine tractor floatplane	Torpedo bomber, production
FF43	Tractor floatplane	Fighter prototype
FF44	Tractor floatplane	Reconnaissance prototype
FF45 (Army Desig.: G.III)	Twin-engine pusher bomber	Bomber, production
FF46 (Army Desig.: D.I)	Tractor landplane	Fighter prototype
FF48	Tractor floatplane	Two-seat fighter prototype
FF49	Tractor floatplane	Reconnaissance, production
FF53	Twin-engine tractor floatplane	Torpedo bomber, prototype
FF54	Tractor quadraplane	Fighter prototype
FF55 (Army Desig.: G.V)	Twin-engine tractor bomber	Bomber, prototype
FF59	Tractor floatplane	Reconnaissance prototype
FF60	Four-engine tractor triplane floatplane	Reconnaissance prototype
FF61 (Army Desig.: G.IV)	Twin-engine pusher bomber	Bomber, production
FF62	Twin-engine pusher bomber	Bomber, prototype
FF63	Tractor floatplane	Two-seat fighter prototype
FF64	Tractor floatplane	Reconnaissance prototype
FF67	Postwar tractor floatplane	Transport prototype
FF71	Postwar tractor floatplane	Transport prototype
Army Desig.: C.I	Tractor landplane (FF Desig. unknown)	Reconnaissance prototype
Army Desig.: N.I	Tractor landplane (FF Desig. unknown)	Night bomber prototype

Above: Manzell on Lake Constance was a key Friedrichshafen facility. Shown here in 1920, the below map gives its location.

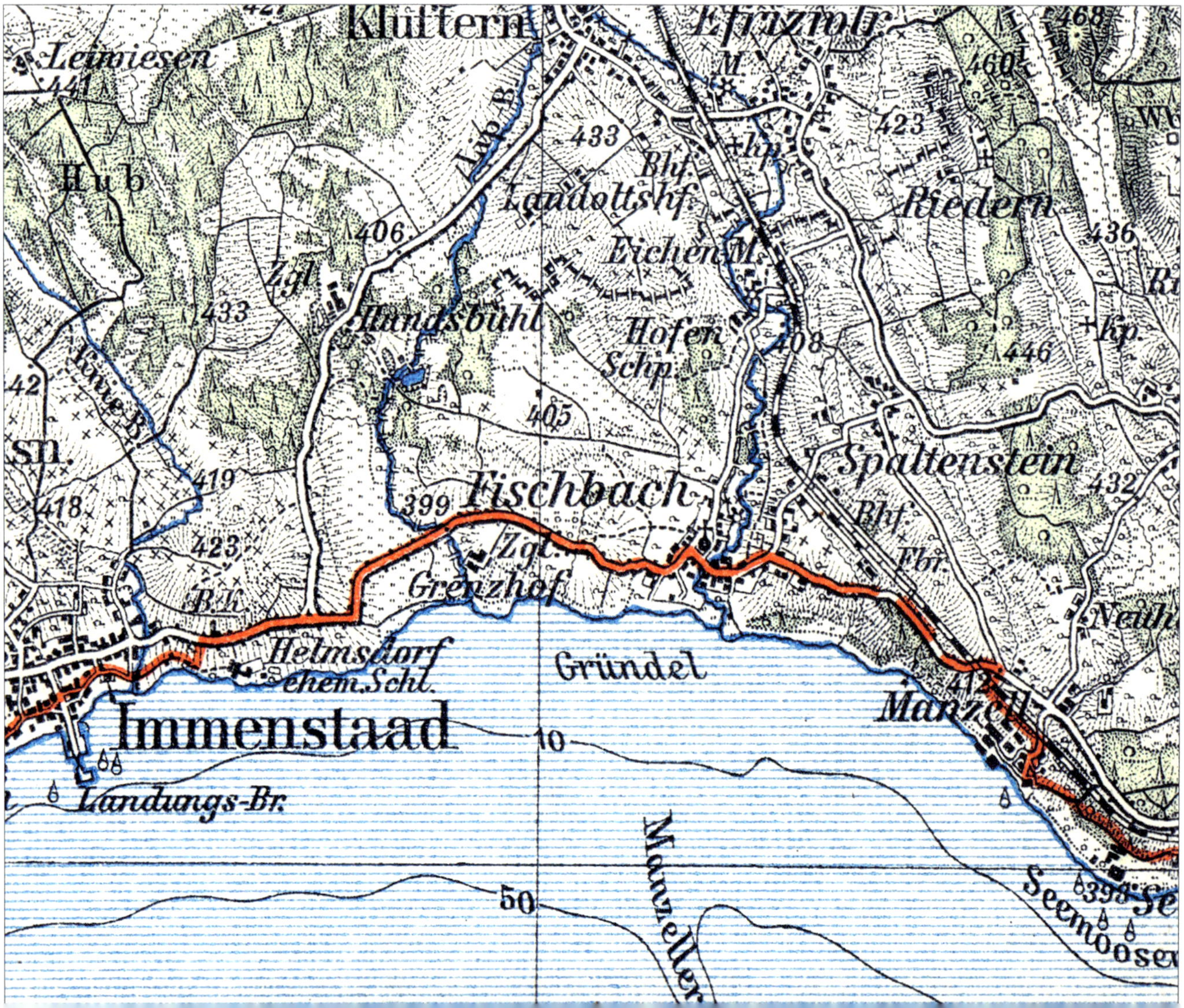

Above: Friedrichshafen G.II(Daim) 625/16 was the first Friedrichshafen bomber built by Daimler, which produced 298 Friedrichshafen bombers under license. Here it is undergoing its type test at Adlershof in April 1917. The G.II was the first Friedrichshafen twin-engine bomber design to see production and operational service and established the basic pattern followed by all subsequent Friedrichshafen bomber designs.

Frontbeststand Inventory of Friedrichshafen Twin-Engine Bombers at the Front

	1914			1915						1916						1917						1918			
Manufacturer and Type	31 Aug	31 Oct	31 Dec	28 Feb	30 Apr	30 Jun	31 Aug	31 Oct	31 Dec	28 Feb	30 Apr	30 Jun	31 Aug	31 Oct	31 Dec	28 Feb	30 Apr	30 Jun	31 Aug	31 Oct	31 Dec	28 Feb	30 Apr	30 Jun	31 Aug
Friedrichshafen G.II														1		4	8	17	17	10	9	2	1	1	1
G.III																		9	32	24	57	69	96	74	24
G.IIIa																								18	95
G.IV																								4	8
G.IVa																								5	6
Friedrichshafen G Total														1		4	8	26	49	34	66	71	97	102	134

Above: Friedrichshafen bombers became increasingly numerous after their introduction and by the end of the war outnumbered bombers from all competing manufacturers. After its introduction the G.IIIa rapidly replaced the earlier G.III.

Below & Below Right: Grave of Theodor Kober and his wife.

SVK Table of Friedrichshafen Seaplane Orders and Deliveries (Table 1 of 3)

Order Number	Type	Marine Numbers	Design	Class & Engine	1914												1915											
					J	F	M	A	M	J	J	A	S	O	N	D	J	F	M	A	M	J	J	A	S	O	N	D
1		25/29	19	100M																								
2		41		120Nag	Not Clarified																							
3		62	27	135Nag																								
4		76/85	19	B 100M								(10)	3	2	5													
5		91/100	29A	B 100M											(10)	2	2	3	1				1	1				
6	117	117	34	CFT 240Mb											(1)		1											
7		201/220	29	B 120M								(20)			4	3	1	3	1	5	1	1	1					
8	274	274/275	31	CB 160Mb								(2)									1			1				
9		290/296	29	B 120A											(7)					1			3	2	1			
10	300	300	35	CBFT 2x160M														(1)										
11		406/410	29	B 120M											(5)	2		2	1									
12		411/415	29A	B 100M											(5)				2	3 Aircraft Not Delivered								
13		416/420	33A	100M												(5)			3			1		1				
14		421	33A	B 100M												(1)								1				
15		426	33	B 120M											(1)					1								
16		451/455	33	B 120M													(5)				2	1	1	1				
17		456/460	33B	CB 160Mb													(5)				2	1	1	1				
18		471/473	33E	BFT 160M															(3)	1	1	1						
19		478/483	33E	BFT 150B																		(6)		2	3			
20		491/496	33E	BFT 150B																		(6)			3	3		
21		497/508	33E	B 150B																				(12)		5	5	2
22		510	33E	C 150B																				(1)		1		
23		529/533	33E	B 150B																						(5)		4
24		534/538	33F	C 150B																						(5)		1
25		541/550	33E	BFT 150B																							(10)	
26		553/562	33E	BFT 150B																							(10)	
27	587	587	39	CFT 200B																								(1)
28	691	591/595	33E	BFT 150B																								
29	745	596/600	33H	C 150B																								
30	691	633/637	33E	BFT 150B																								
31	745	638/642	33H	C 150B																								
32	643/645	643/645	39	CHFT 200B																								
33	691	659/663	33E	BFT 150B																								
34	745	664/668	33H	C 150B																								
35	669	669	40	CFT 240Mb	2 Propeller																							
36	678	678	41A	T 2x150B																								
37	691	680/684	33E	BFT 150B																								
38	745	685/689	33H	C 150B																								
39	691	690/694	33E	BFT 150B																								
40	745	695/699	33H	C 150B																								
41	718	715/724	33E	BFT 150B																								
42	718	725/729	33E	HFT 150B																								
43	745	730/734	33H	C 150B																								
44		735	—	T 2x220B																								
45	718	736/740	33E	HFT 150B																								
46	745	741/745	33H	C 150B																								
47	749	749	43	ED 160M																								
48	718	802/811	33E	HFT 150B																								
49	819	812/821	33H	C 150B																								
50	829	822/841	33E	HFT 150B																								
51	829	850/859	33E	HFT 150B																								
52	829	860/869	33E	B 150B																								
53	876	870/881	33E	HFT 150B																								
54	876	882/884	33E	HFT 150B																								
55	876	885/889	33E	B 150B																								
Orders that Month												32			29	6	10	1	3			12		13		10	20	1
Deliveries that Month													3	2	9	7	3	8	8	8	7	5	7	10	7	9	5	7
Orders that Year																67												70
Orders: (3) Deliveries: 2 — Deliveries that Year																21												84

Orders are shown as circled numbers, written here as (N); deliveries are shown as plain numbers.

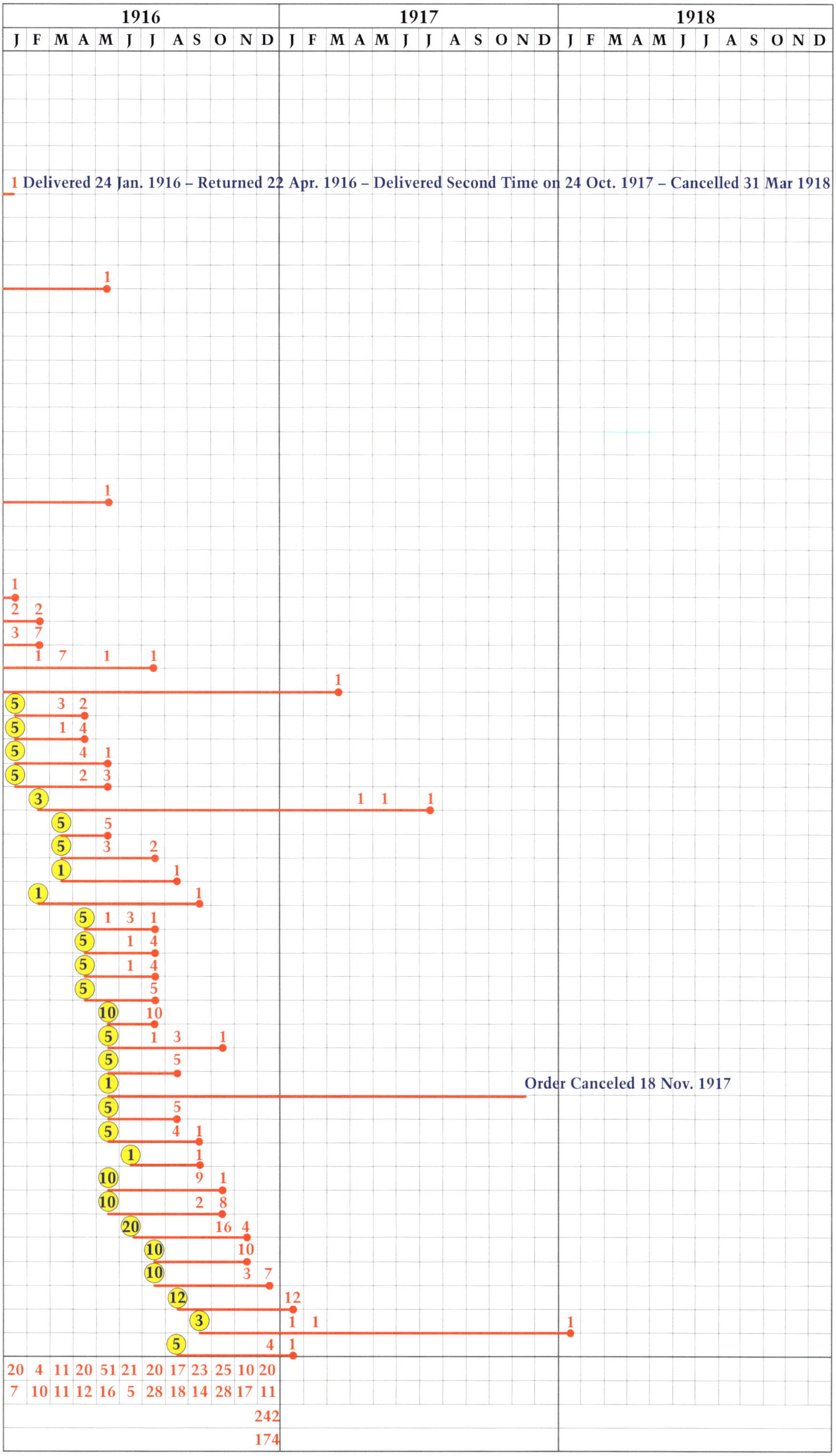
1916
1917
1918
J F M A M J J A S O N D
1 Delivered 24 Jan. 1916 – Returned 22 Apr. 1916 – Delivered Second Time on 24 Oct. 1917 – Cancelled 31 Mar 1918
Order Canceled 18 Nov. 1917
20 4 11 20 51 21 20 17 23 25 10 20
7 10 11 12 16 5 28 18 14 28 17 11
242
174

SVK Table of Friedrichshafen Seaplane Orders and Deliveries (Table 2 of 3)

Order Number	Type	Marine Numbers	Design	Class & Engine	1916												1917													Notes
					J	F	M	A	M	J	J	A	S	O	N	D	J	F	M	A	M	J	J	A	S	O	N	D	J	
6	117	117	34	CFT 240Mb	1																									Delivered 24 Jan. 1916 – Returned 22 Apr. 1916 – Delivered Second Time on 24
10	300	300	35	CBFT 2x160M					1																					
19		478/483	33E	BFT 150B					1																					
23		529/533	33E	B 150B	1																									
24		534/538	33F	C 150B	2	2																								
25		541/550	33E	BFT 150B	3	7																								
26		553/562	33E	BFT 150B		1	7		1		1																			
27	587	587	39	CFT 200B															1											
28	691	591/595	33E	BFT 150B	(5)		3	2																						
29	745	596/600	33H	C 150B	(5)		1	4																						
30	691	633/637	33E	BFT 150B	(5)			4	1																					
31	745	638/642	33H	C 150B	(5)			2	3																					
32	643/645	643/645	39	CHFT 200B		(3)														1	1		1							
33	691	659/663	33E	BFT 150B			(5)		5																					
34	745	664/668	33H	C 150B			(5)		3	2																				
35	669	669	40	CFT 240Mb			(1)				1																			2 Propeller
36	678	678	41A	T 2x150B		(1)						1																		
37	691	680/684	33E	BFT 150B				(5)	1	3	1																			
38	745	685/689	33H	C 150B				(5)		1	4																			
39	691	690/694	33E	BFT 150B				(5)		1	4																			
40	745	695/699	33H	B 150B				(5)			5																			
41	718	715/724	33E	BFT 150B					(10)		10																			
42	718	725/729	33E	HFT 150B					(5)		1	3		1																
43	745	730/734	33H	C 150B					(5)			5																		
44		735		T 2x220B					(1)																					Order Canceled 18 November 1917
45	718	736/740	33E	BFT 150B					(5)			5																		
46	745	741/745	33H	C 150B					(5)			4	1																	
47	749	749	43	ED 160M						(1)			1																	
48	718	802/811	33E	BFT 150B					(10)				9	1																
49	819	812/821	33H	C 150B					(10)				2	8																
50	829	822/841	33E	HFT 150B						(20)				16	4															
51	829	850/859	33E	HFT 150B							(10)				10															
52	829	860/869	33E	B 150B							(10)				3	7														
53	876	870/881	33E	HFT 150B								(12)					12													
54	876	882/884	33E	HFT 150B									(3)				1	1											1	
55	876	885/889	33E	B 150B								(5)				4	1													
56		922/929	33J	HFT 150B									(8)					2	5	1										
57		930/931	33J	HFT 150B									(2)					1		1										
58	933	932/941	33L	C2MG 150B									(10)				2	8												
59	937	996/1000	41A	T 2x150B										(5)							3	1	1							
60	1004	1001/1010	33L	CHFT 150B											(10)				6	3			1							
61	1004	1085/1094	33L	CHFT 150B										(10)						7	2	1								
62		1095/1104	33J	HTF 150B										(10)					1	1	5	3								
63	1004	1117/1126	33L	C2MG 150B												(10)				1	5	4								
64	1127	1127/1136	39	CHFT 200B												(10)							2	7	1					
65	1004	1158/1177	33L	CHFT 150B													(20)			12	6	2								
66	997	1208/1210	41A	T 2x150B													(3)						2	1						
67	1004	1234/1278	33L	CHFT 150B														(45)			12	13	17		2	1				
68	1004	1279/1288	33L	C2MG 150B														(10)				10								
69		1289/1298	33J	HFT 150B														(10)					2	8						
70	1472	1472/1474	48	C2MG 240Mb																(3)						3				
71	1699	1521/1535	49C	CHFT 200B																	(15)				9	6				
72	1549	1536/1550	49B	BFT 200B																	(15)				1	14				
73	1583	1577/1596	33L	C2MG 150B																		(20)		7	13					
				Orders that Month	20	4	11	20	51	21	20	17	23	25	10	20	23	65	—	3	30	20	90	120	30	—	—	30	—	
				Deliveries that Month	7	10	11	12	16	7	27	18	13	26	17	11	16	12	13	27	34	34	26	23	26	36	15	?	?	
				Orders that Year												242														
Orders: (3) Deliveries: 2				Deliveries that Year												175														

SVK Table of Friedrichshafen Seaplane Orders and Deliveries (Table 3 of 3)

Order Number	Type	Marine Numbers	Design	Class & Engine	1917												1918											
					J	F	M	A	M	J	J	A	S	O	N	D	J	F	M	A	M	J	J	A	S	O	N	D
27	587	587	39	CFT 200B			1																					
32	643/645	643/645	39	CHFT 200B				1	1		1																	
44		735		T 2x220B											Order Canceled 18 November 1917													
53	876	870/881	33E	HFT 150B	12																							
54	876	882/884	33E	HFT 150B	1	1											1											
55	876	885/889	33E	B 150B	1																							
56		992/929	33J	HFT 150B		2	5	1																				
57		930/931	33J	HFT 150B		1		1																				
58	933	932/941	33L	C2MG 150B	2	8																						
59	997	996/1000	41A	T 2x150B					3	1	1																	
60	1004	1001/1010	33L	CHFT 150B			6	3			1																	
61	1004	1085/1094	33L	CHFT 150B				7	2	1																		
62		1095/1104	33J	HFT 150B			1	1	5	3																		
63	1004	1117/1126	33L	C2MG 150B				1	5	4																		
64	1127	1127/1136	39	CHFT 200B							2	7	1															
65	1004	1158/1177	33L	CHFT 150B	(20)			12	6	2																		
66	997	1208/1210	41A	T 2x150B	(3)						2	1																
67	1004	680/684	33E	BFT 150B		(45)			12	13	17		2	1														
68	1004	1234/1278	33L	C2MG 150B		(10)				10																		
69		1289/1298	33J	HFT 150B		(10)					2	8																
70	1472	1472/1474	48	C2MG 240Mb				(3)						3														
71	1699	1521/1535	49C	CHFT 200B					(15)				9	6														
72	1549	1536/1550	49B	BFT 200B					(15)				1	14														
73	1583	1577/1596	33L	C2MG 150B						(20)		7	13															
74	1699	1597/1606	49C	CHFT 200B						(10)				9		1												
75	1549	1607/1609	49B	CHFT 200B						(3)				3														
76	1549	1610/1616	49B	BFT 200B						(7)					6	1												
77	1663	1663/1665	53	T 2x260M						(3)																		
78	1699	1669/1698	49C	CHFT 200B							(30)				9	19	1	1										
79	1699	1699/1718	49C	CHFT 200B								(20)				7	12	1										
80	1699	1742/1811	49C	CHFT 200B								(70)					16	32	22									
81	1822	1812/1841	49C	C2MGHFT 200B								(30)					1	2	3	6	7	1						
82	3025	1980/1999	33E	S 150B									(20)				2	18										
83	1822	2053/2092	49C	CHFT 200B									(40)															
84	3025	3001/3030	33S	S 150B												(30)		6	15		1	5						
85		3031/3060	33J	HFT 150B															(30)									
86		3061/3063	64	CHFT 160M															(3)									
87		3064/3083	49C	CHFT 200B																		(20)						
88		3301/3302	60	C3MG 4x160M	Long-Range Reconnaissance Plane													(2)										
89		3303	60	C3MG 4x160M	Long-Range Reconnaissance Plane																	(1)						
				Orders that Month	20	4	—	3	30	43	30	120	60	—	—	20	—	2	33	—	—	21						
				Deliveries that Month	18	12	13	27	34	7	27	18	13	26	17	11	33	30	40	6	8	6						
				Orders that Year												404												
Orders: (3)	Deliveries: 2			Deliveries that Year												288												

Above, Facing Page, Previous Spread: Friedrichshafen produced more floatplanes for the German Navy than any other company, and the *SVK* tables of Friedrichshafen orders and deliveries covered three pages. These tables are reproduced here in the same format as the originals but recreated for legibility. Unfortunately, no data was added to these tables after June 1918.

Pre-War Friedrichshafen Aircraft

Flugzeugbau Friedrichshafen GmbH was founded in 1912 and its first aircraft, the FF1, was built and flown that same year. Friedrichshafen went on to build a surprising number of new aircraft designs before the war started, reflecting both Friedrichshafen's dedication to their business and the simplicity and ease of construction of the primitive aircraft of the time. Most of Friedrichshafen's new designs were seaplanes, and most were limited to a single prototype.

As Friedrichshafen progressed along the technological learning curve, they were able to build new designs that warranted limited production. The outbreak of war greatly increased demand for practical military aircraft, and Friedrichshafen's production orders increased accordingly.

Friedrichshafen FF1

As indicated by its factory designation, the FF1 was the first Friedrichshafen aircraft to be built. First flight of this two-seat floatplane was in mid-November 1912. The FF1 was a biplane pusher powered by an 100 hp Argus engine, and wheels could be attached for ground handling. Initially the span of upper and lowers wings was 14.0 m and 10.0 m respectively, but later the span was enlarged to 16.8 m and 14.0 m respectively. On Feb. 14, 1913, pilot Gsell set a new flight endurance record with the FF1 by flying for 2 hours, 32 minutes, and 30 second. In February 1914 the sole FF1 built was destroyed in a crash.

Friedrichshafen FF1 Specifications

Engine:	100 hp Argus	
Wing:	Span Upper	14.00 m
	Span Lower	11.00 m
	Area	48.0 m^2
General:	Length	11.15 m
	Height	3.35 m
	Empty Weight	1,672 kg
	Loaded Weight	1,972 kg
Maximum Speed:		85 km/h
Climb:	560 m	8 min

Below: Friedrichshafen FF1 with Dr. Köster handling the large pilot's control wheel.

Above: The Friedrichshafen FF1 had a central main float that could be fitted with wheels to facilitate ground handling. Side-mounted radiators cooled the Argus engine and the ailerons were mounted between the wings.

Below: The Friedrichshafen FF1 could carry two crewmen, with the pilot in the front cockpit.

Friedrichshafen FF2

In 1913 the second Friedrichshafen design, the FF2, made its first flight. It too was a floatplane, but had a totally different configuration than the FF1, being a tractor monoplane. A single-seater, it was smaller and less powerful than the FF1, having a 50 hp Orlikon engine. The FF2 was derived from a monoplane design by the Swiss designer Grandjean, who sold his patent to Friedrichshafen. The FF2 was built especially for the customer Ehrler from Unteruhldingen on Lake Constance. Only one FF2 was built.

Friedrichshafen FF2 Specifications		
Engine:	50 hp Orlikon	
Wing:	Span	11.00 m
	Area	48.0 m^2
General:	Length	9.80 m
	Loaded Weight	972 kg

Below: Built, this for a private customer, only one FF2, the second Friedrichshafen design, was constructed.

Above: The Friedrichshafen FF2 was a small, single-seat monoplane with two floats that was built for a private customer.

Early production FF33E floatplanes had a tail float and a rudder of larger chord than later production FF33E aircraft. In addition, MN 501 uses the early side-mounted radiator.

Friedrichshafen FF4

Apparently the FF3 design was not built, because the FF4 was the next aircraft from Friedrichshafen. The FF4 was a development of the FF2 and was built to train Ehrler, who did not compete in the 1914 Lake Competition. First flying in early February 1914, the FF4 was a more mature two-seat floatplane of tractor monoplane configuration that loosely followed the basic *Taube* (dove) configuration so popular in Germany at the time for its inherent stability. It was powered by a four-cylinder, 65 hp Argus. Little is known about its specifications, and the sole FF4 burned during a major fire in 1918 in the airship hangar.

Friedrichshafen FF4 Specifications	
Engine:	65 hp Argus

Above & Below: Two views of the Friedrichshafen FF4 training airplane built for a private customer. (Lower photo the Peter M. Bowers Collection/The Museum of Flight)

Friedrichshafen FF6

The FF5 design was not built, making the next Friedrichshafen aircraft the FF6, the first Friedrichshafen land plane. The FF6 was a two-seat tractor biplane with a 2½ bay wing cellule powered by an in-line, water-cooled engine, perhaps a 100 hp Argus. The FF6 was a conventional design with ailerons for roll control and one photo shows it with serial B203 on the rudder, indicating it may have been used for reconnaissance or training early in the war. As far as is known, only a single aircraft was built and no specifications have survived.

Friedrichshafen FF6 Specifications	
Engine:	100 hp Argus?

Above & Below: Two views of the Friedrichshafen FF6 two-seat land plane. The tail markings (B203) below indicate the possibility it flew reconnaissance missions early in the war. (The Peter M. Bowers Collection/The Museum of Flight)

Friedrichshafen FF8

The FF7 design was not built and with their next design, the FF8, Friedrichshafen returned to a floatplane of tractor monoplane configuration. The first flight of the FF8 was in June 1913; like the earlier FF4 it used a 65 hp, four-cylinder Argus for power. Wing warping was used for roll control, but the wing did not use the *Taube* planform, which probably improved its performance compared to the FF4. Again only one aircraft was built.

Friedrichshafen FF8 Specifications		
Engine:	65 hp Argus	
Wing:	Span	13.00 m
	Area	22.00 m^2
General:	Length	11.00 m
	Height	3.60 m
	Loaded Weight	620 kg
Maximum Speed:		95 km/h

Left: The Friedrichshafen FF8 floatplane before markings were applied. The FF8 reverted to wing-warping for roll control. One reference states the FF8 was a single-seater, although it clearly had room for a second seat in front of the pilot and it seems a passenger could have been carried. (The Peter M. Bowers Collection/ The Museum of Flight)

Below: The Friedrichshafen FF8 floatplane after markings were applied. The ground crew appear to be in the process of launching the aircraft.

Friedrichshafen FF9

With the FF9 that first flew in June 1913, Friedrichshafen returned to the pusher configuration. The FF9 strongly resembled the earlier FF1 but was larger and more powerful, using a 135 hp NAG six-cylinder engine. As with the FF1, wheels could be attached to simplify ground handling.

Pilot Dahms flew a long distance course from Manzell over Basel-Cologne-Amsterdam-Emden to Hamburg in the FF9 during the period 20–23 September 1913. However, in Blankensee the plane crashed and was badly damaged, Dahms and his passenger being injured. The FF9 was repaired and on 12 November was flown to a new world endurance record for biplanes, flying for 6 hours, 16 minutes, and 30 seconds. Like earlier Friedrichshafen aircraft, only a single example was built.

Friedrichshafen FF9 Specifications		
Engine:	135 hp NAG	
Wing:	Span Upper	15.10 m
	Span Lower	10.35 m
	Area	42.0 m^2
General:	Length	12.00 m
	Empty Weight	850 kg
	Loaded Weight	1,250 kg
Maximum Speed:		115 km/h

Above: The Friedrichshafen FF9 reverted to the configuration of the earlier FF1 and greatly resembled it, but was a larger, more powerful aircraft.

Left: *Obermeister* Sauter in the cockpit of the Friedrichshafen FF9.

MN 559 represents a transition between early and later production FF33E floatplanes; it had the early large-chord rudder but the tail float was replaced by a small fin. The fuselage cross had a thin black outline but the tail and wing crosses are on white backgrounds.

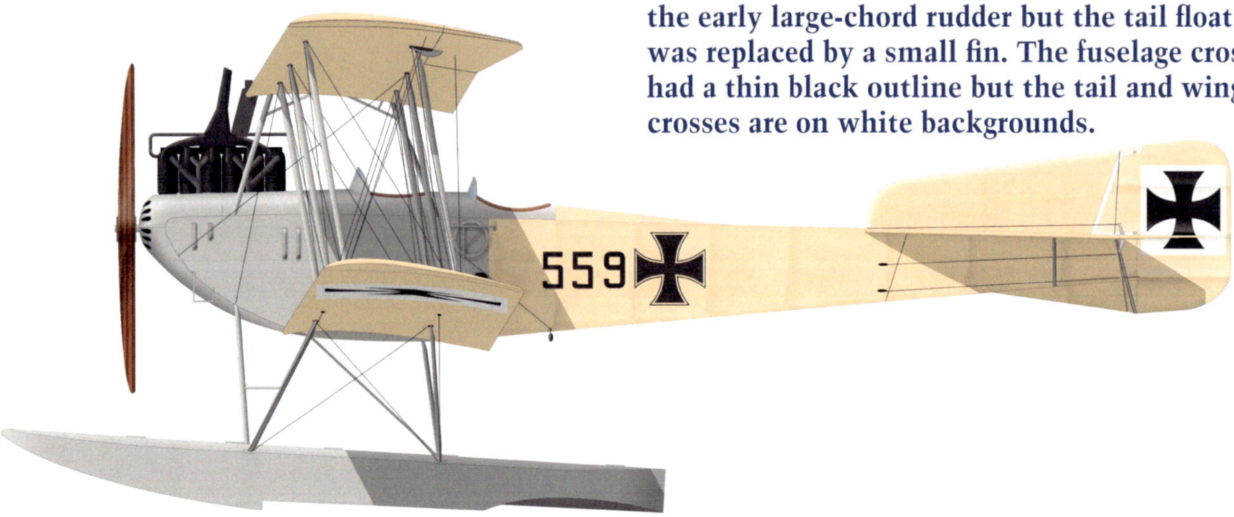

Above: The Friedrichshafen FF9 was a larger, more powerful aircraft than the FF1 it resembled. Here it uses the wheels that could be extended for land operations. The text on the rudder reads "F.F. Flugzeugbau Friedrichshafen G.m.b.H."

Above: The Friedrichshafen FF9 attracts attention in Düsseldorf. All its structural details are exposed for examination.

Above: With its wheels folded out of the way the Friedrichshafen FF9 operates from water at Hamburg, 1913.

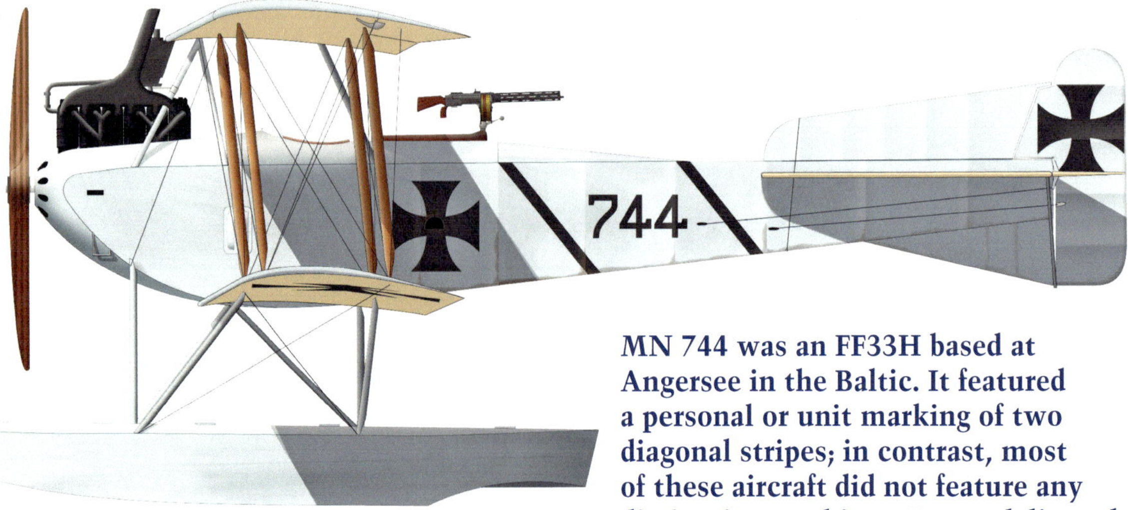

MN 744 was an FF33H based at Angersee in the Baltic. It featured a personal or unit marking of two diagonal stripes; in contrast, most of these aircraft did not feature any distinctive markings. It was delivered in August or September 1916.

Friedrichshafen FF11

The FF10 was another unbuilt design, and next Friedrichshafen aircraft to appear was the FF11, the first Friedrichshafen flying boat. Like the FF9 it was powered by a 135 hp NAG six-cylinder engine, but this was mounted in the hull and drove a pusher propeller via a long drive shaft, enabling the propeller to be raised above the spray thrown up by the hull during take-off and landing. The Oertz yacht works built the hull for the FF11 at Reiherstieg in Hamburg; subsequent Oertz designs benefited from this connection as can be seen by looking at the Oertz flying boat designs.

First flight of the FF11 was in February 1914 and it was tested until the end of May on the Bodensee, then shipped to the SVK on 4 June 1914, receiving Marine Number 41. Like most other pre-war Friedrichshafen aircraft, only one FF11 was built.

Friedrichshafen FF1 Specifications

Engine:	135 hp NAG	
Wing:	Area	53.0 m^2
General:	Length	11.15 m
	Height	3.35 m
	Empty Weight	1,180 kg
	Loaded Weight	1,500 kg
Maximum Speed:		90 km/h
Climb:	500 m	15 min

Above & Right: The FF11 was Friedrichshafen's first flying boat. The hull was built by the Oertz yacht works and the similarity to Oertz flying boats is obvious. The similarly-powered FF9 had notably better performance than the FF11 due to the flying boat's greater size and weight. The SVK assigned Marine Number 41 to the FF11.

Left: WWI advertisement for Mercedes engines for automobiles and airplanes.

Friedrichshafen FF15

The next Friedrichshafen aircraft design to be built was the FF15, an aircraft that owed much to the earlier FF1 and FF9. The FF15 was an improved FF9 and used the same 135 hp NAG as the FF9; other specifications were very similar.

Pilot Truckenbrodt flew the sole FF15 at the 1914 Bodensee Competition.

Friedrichshafen FF15 Specifications		
Engine:	135 hp NAG	
Wing:	Span Upper	15.10 m
	Area	42.0 m²
General:	Length	12.00 m
	Empty Weight	850 kg
	Loaded Weight	1,270 kg
Maximum Speed:		115 km/h

This Page: The Friedrichshafen FF15 followed the configuration of the earlier FF1 and FF9. The FF15 was basically an improved FF9 and had similar specifications and engine type.

Friedrichshafen FF17

The FF17 floatplane was the next Friedrichshafen design built. It was a tractor, two-seat biplane with 2½ bay wing cellule. Powered by a 135 hp NAG six-cylinder engine, it was the first biplane with full fuselage built by Friedrichshafen, and its first flight was in May 1914. It was built for and flown at the 1914 Bodensee competition. At first it used a central float with wingtip floats for balance, but then was converted to a pair of floats. The intention was to fly the modified aircraft in the 1914 Baltic flight competition. Like the earlier Friedrichshafen aircraft, only one was built, but it had the potential to be a reconnaissance floatplane.

Friedrichshafen FF1 Specifications		
Engine:	135 hp NAG	
Wing:	Span Upper	15.20 m
	Span Lower	10.20 m
General:	Length	10.00 m

Above: The FF17 was a much more modern design than earlier Friedrichshafen seaplanes. Here the FF17A has a single central float with large wingtip floats for stability on the water.

Below: The FF17 was later tested with dual main floats as the FF17B. This configuration resembled the Friedrichshafen seaplanes used during the war. (The Peter M. Bowers Collection/The Museum of Flight)

Friedrichshafen FF19

Although a later design, the FF19 actually flew in April 1914 before the FF17. It featured a three-bay biplane wing cellule that was simpler than the FF17, perhaps one reason it was flown earlier. A tractor floatplane powered by a 100 hp Mercedes D.I, it was the first Friedrichshafen pre-war design to go into limited production, a total of 16 being built.

The first FF19 order for five aircraft, Marine Numbers 25–29, was built in the period 17 May to 18 June 1914 in Kiel. The second order was for 10 aircraft, Marine Numbers 76–85, immediately after the outbreak of war. The FF19s, together with 3 Agos ordered at the same time, were then the only naval maritime aircraft capable of fulfilling a combat role. The 16th airframe was used for static load testing.

Friedrichshafen FF19 Specifications		
Engine:	100 hp Mercedes D.I	
Wing:	Span Upper	15.30 m
	Span Lower	13.80 m
	Area	48.0 m^2
General:	Length	10.40 m
	Empty Weight	860 kg
	Loaded Weight	1,160 kg
Maximum Speed:		95 km/h
Climb:	500 m	8 min

The FF19 was historically significant as the first German aircraft from which successful wireless tests were carried out.

Above: The FF19 was the first Friedrichshafen design to be produced in quantity. A total of 16 airframes were built; 15 were delivered to the Navy and one airframe was used for structural testing.

Friedrichshafen FF19 Production		
Marine Numbers	Quantity	Date Ordered
25–29	5	Unknown
76–85	10	Aug. 1914

Left: The first small production batch of FF19s was ordered pre-war; the outbreak of war motivated the second order.

Above: The first production Friedrichshafen FF19, Marine Number 25, trails a red wingtip identification streamer.

Below: A Friedrichshafen FF19 taxis while another flies overhead.

Above: Friedrichshafen FF19 Marine Number 25 at Borkum wearing late-style insignia. Research by Frits Gerdessen reveals that, on 14 August 1914, Friedrichshafen FF19 seaplane #28 beached at Schiermonnikoog. The plane was later acquired for the Netherlands Navy and the crew, *Oberleutnant zur See* Hubert Waldemar Klein and *Fähnrich zur See* Carl Georg Philipp, was interned at Bergen. This was the first of many naval aircraft interned by the Netherlands during the war.

Above: Friedrichshafen FF19 Marine Number 78 at Borkum.

Friedrichshafen FF21

The FF21 was an improved version of the Friedrichshafen FF11 developed by Jaray especially for the 1914 Baltic Seaplane Competition. The 150 hp Benz Bz.III engine was built into the flying boat hull and drove the propeller via an extension shaft and bevel gear to keep the four-blade propeller up out of the spray of the hull during landing and take-off.

The flying boat hull was built by the Oertz yacht works, which had also built the FF11 hull, and was clearly related to a motorboat. Again, like most pre-war Friedrichshafen aircraft, only a single FF21 was built.

Friedrichshafen FF21 Specifications		
Engine:	150 hp Benz Bz.III	
Wing:	Span Upper	16.90 m
	Span Lower	13.70 m
General:	Length	11.50 m
	Height	2.90 m

Above & Below: The FF21 was developed from the Friedrichshafen FF11 and looks virtually identical in side view to the Oertz F.B.1, leading one to wonder if they were indeed the same design. Oertz built the hulls for both the FF11 and FF21, and there is a great resemblance between these designs.

Friedrichshafen FF27

The final pre-war Friedrichshafen design was the FF27. The FF27 combined some characteristics of earlier Friedrichshafen designs, being a two-seat pusher biplane floatplane with a three-bay wing cellule that could have wheels attached for improved ground handling. Like several preceding Friedrichshafen aircraft, a 135 hp NAG six-cylinder engine provided the power. Although designed for the 1914 Baltic Seaplane Competition, the sole FF27 built was accepted by the *SVK* and assigned Marine Number 62.

Friedrichshafen FF27 Specifications

Engine:	135 hp NAG	
Wing:	Span Upper	14.40 m
	Span Lower	13.40 m
	Area	45.0 m^2
General:	Empty Weight	900 kg
	Loaded Weight	1,100 kg
Maximum Speed:		100 km/h
Climb:	500 m	8 min

Above: The FF27 was a refinement of earlier Friedrichshafen pusher biplanes; the one built was accepted by the German Navy and assigned Marine Number 62. Wheels could be attached as shown for easier ground handling. The drag-creating pusher configuration was little used by German manufacturers after the war started. Pushers had good visibility for the crew but were too slow. Later, when air-to-air combat became more common, pushers were too vulnerable to attack from the rear.

Friedrichshafen Reconnaissance Floatplanes

Upon the outbreak of war the demand for useful reconnaissance aircraft leaped, a situation as true for naval forces as much as for the armies. Orders for Friedrichshafen reconnaissance floatplanes were no exception, and the aircraft available were improved versions of pre-war designs.

Friedrichshafen FF29 & FF29A

Friedrichshafen's first wartime design was the FF29, which was basically an enlarged FF19, the first Friedrichshafen aircraft that had been successful enough to be built in any quantity. The FF29 and FF29A were built as unarmed, two-seat reconnaissance floatplanes.

Production orders for the 44 aircraft actually built are somewhat complex and are summarized in the adjacent table. The first batch of 20 FF29 aircraft were ordered at the start of the war in August 1914. All subsequent batches were ordered in November 1914 and the final aircraft from this orders was delivered in September 1915. One batch of FF29 aircraft, Marine Numbers 290–296, were ordered with the 120 hp Argus.

Friedrichshafen FF29 Specifications		
Engine:	120 hp Mercedes D.II	
Wing:	Span Upper	16.30 m
	Area	57.5 m^2
General:	Length	10.40 m
	Empty Weight	928 kg
	Loaded Weight	1,400 kg

The FF29A was the designation for aircraft built under license by the *Kaiserliche Werft* Wilhelmshaven. These were powered by the 100 hp Mercedes D.I and had slightly modified floats and tail.

Above: FF29s Marine Numbers 208, 204, and 407 are lined up with Oertz F.B.3 MN 46 at Zeebrügge in early 1915.

Above: Friedrichshafen FF29 Marine Number 211 was an FF29 powered by a 120 hp Mercedes D.II.

Above: Friedrichshafen FF29 #17 in Danish service.

Facing Page: Early on 6 January 1915 an FF29a was lashed to the deck of the U-Boat *U-12* as an experiment to see if the submarine could take the aircraft out into the North Sea where it could perform reconnaissance duties for the submarine. The machine had to be lashed with the wings parallel to the submarine's length. Despite a fairly heavy sea the FF29a was launched successfully about 30 miles out from Zeebrugge. Unfortunately, the conditions deteriorated such that the pilot of the seaplane did not try to make the rendezvous with the submarine but flew back to base. While the idea had worked it was apparent that much more work would be required to develop a true integration of the submarine and an aircraft. This photo may have been related to that experiment. (See: *Submarines with Wings* by T.C. Treadwell, Conway, 1985.)

Friedrichshafen FF29 & FF29A Production				
Type	Marine Numbers	Quantity	Date Ordered	Naval Class & Notes
FF29A	91–100	10	November 1914	B, 100 hp Mercedes D.I
FF29	201–220	20	August 1914	B, 120 hp Mercedes D.II
FF29	290–296	7	November 1914	B, 120 hp Argus
FF29	406–410	5	November 1914	B, 120 hp Mercedes D.II
FF29A	411–415	5	November 1914	B, 100 hp Mercedes, 3 aircraft not delivered

Above: Friedrichshafen FF29 Marine Number 2?3 *JUNGDEUTSCHLAND* displays a different tail design than the other FF29 aircraft shown here.

Right: Friedrichshafen FF29 Marine Number 204 and submarine *U-12*.

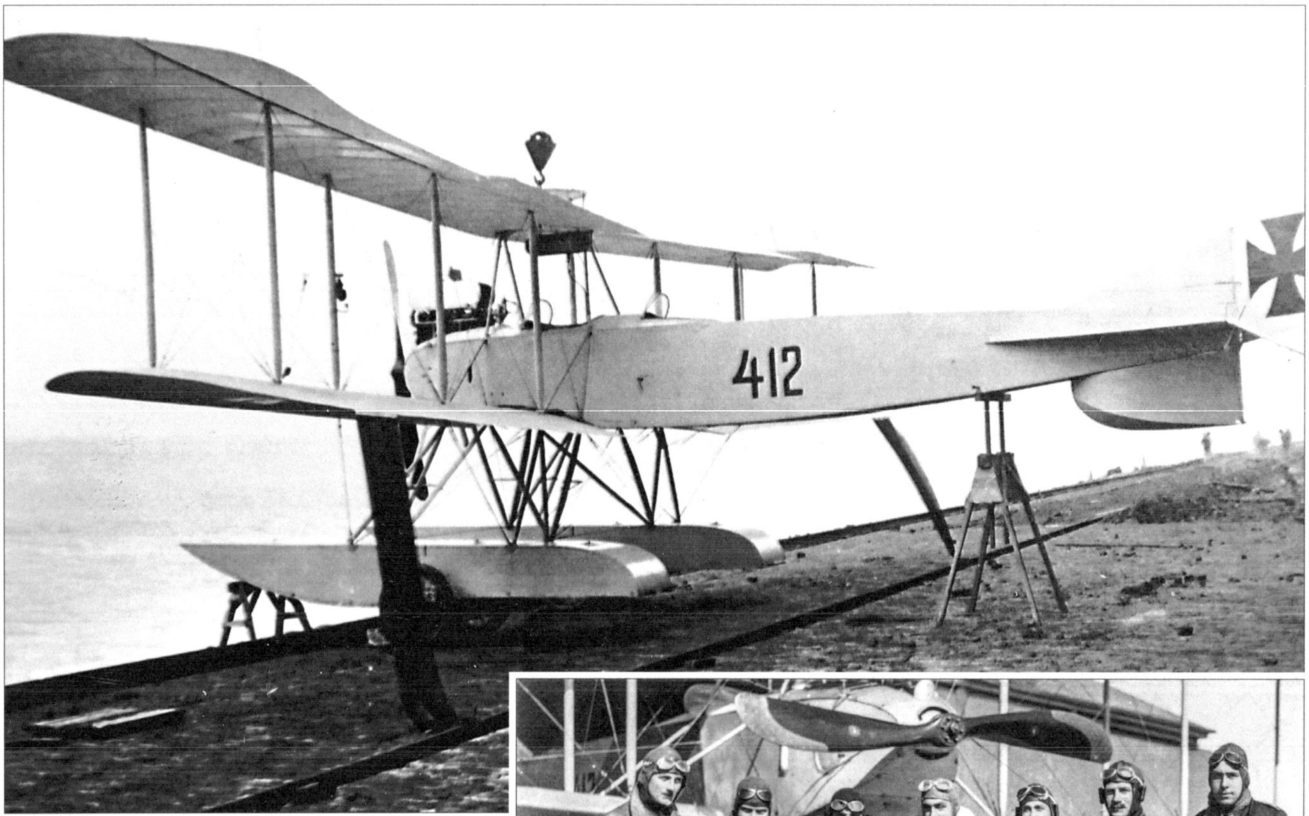

Above: Friedrichshafen FF29A Marine Number 412 was delivered in March 1915.

Right: Friedrichshafen FF29A Marine Number 412 at Borkum. In the center is *Flugmaat* Elsässer, who provided these Borkum photos.

Below: A Friedrichshafen FF29 being brought ashore.

Above: Friedrichshafen FF29 Marine Number 410 was delivered in March 1915. Powered by a 120 hp Mercedes D.II, it was the last aircraft of the last batch of FF29 aircraft ordered.

Below: Friedrichshafen FF29A Marine Number 413. Powered by a 100 hp Mercedes D.I, it was from the last batch of FF29A aircraft ordered.

Friedrichshafen FF31

Above: Friedrichshafen FF31 Marine Number 274 was one of only two FF31 aircraft built, the other being MN 275.

The outbreak of war made arming aircraft imperative, and in the absence of a reliable synchronizing gear at the time, an obvious solution was the pusher configuration. This provided a gunner with a flexible machine gun in the front cockpit an unexcelled field of fire forward without the problem of safely firing the gun through the propeller arc. Moreover, at this time, before any experience with air-to-air combat, the precedent for aerial combat was naval combat between ships with rotating gun mounts, so lack of a synchronizer was not considered a serious drawback if it was considered at all.

Friedrichshafen's approach to this problem for seaplanes was the FF31, two prototypes of which were ordered in August 1914. Their naval class designation was CB, indicating a two-seat reconnaissance airplane with machine gun armament; later this would be simplified to class C. These two aircraft, Marine Numbers 274 and 275, were delivered in May and August 1915.

The FF31 built on the experience of the earlier FF27 design. To expedite these aircraft, the wings and floats of the FF29 were used. Pushers had more drag than equivalent tractor designs with conventional fuselages, and the FF31 was fitted with a 160 hp Maybach, which had notably more power then the 120 hp engines used by the FF29 then in production.

The two FF31 prototypes were thoroughly evaluated. Performance was thought satisfactory for a reconnaissance type, but the speed and climb were certainly not adequate for an interceptor. In addition, the pusher configuration was not favored due to its vulnerability to attack from the rear, and no further FF31 aircraft were ordered.

Friedrichshafen FF31 Specifications

Engine:	160 hp Maybach Mb.III	
Wing:	Span Upper	16.85 m
	Span Lower	15.80 m
	Area	60.0 m^2
General:	Length	10.15 m
	Height	3.70 m
	Empty Weight	1,063 kg
	Loaded Weight	1,535 kg
Maximum Speed:		98 km/h
Climb:	500 m	9.5 min
	800 m	16 min
Range:		500 km

Above & Below: Friedrichshafen FF31 Marine Number 275.

Right: WWI advertisement for Benz automobile and aircraft engines. Most German naval aircraft used Benz engines.

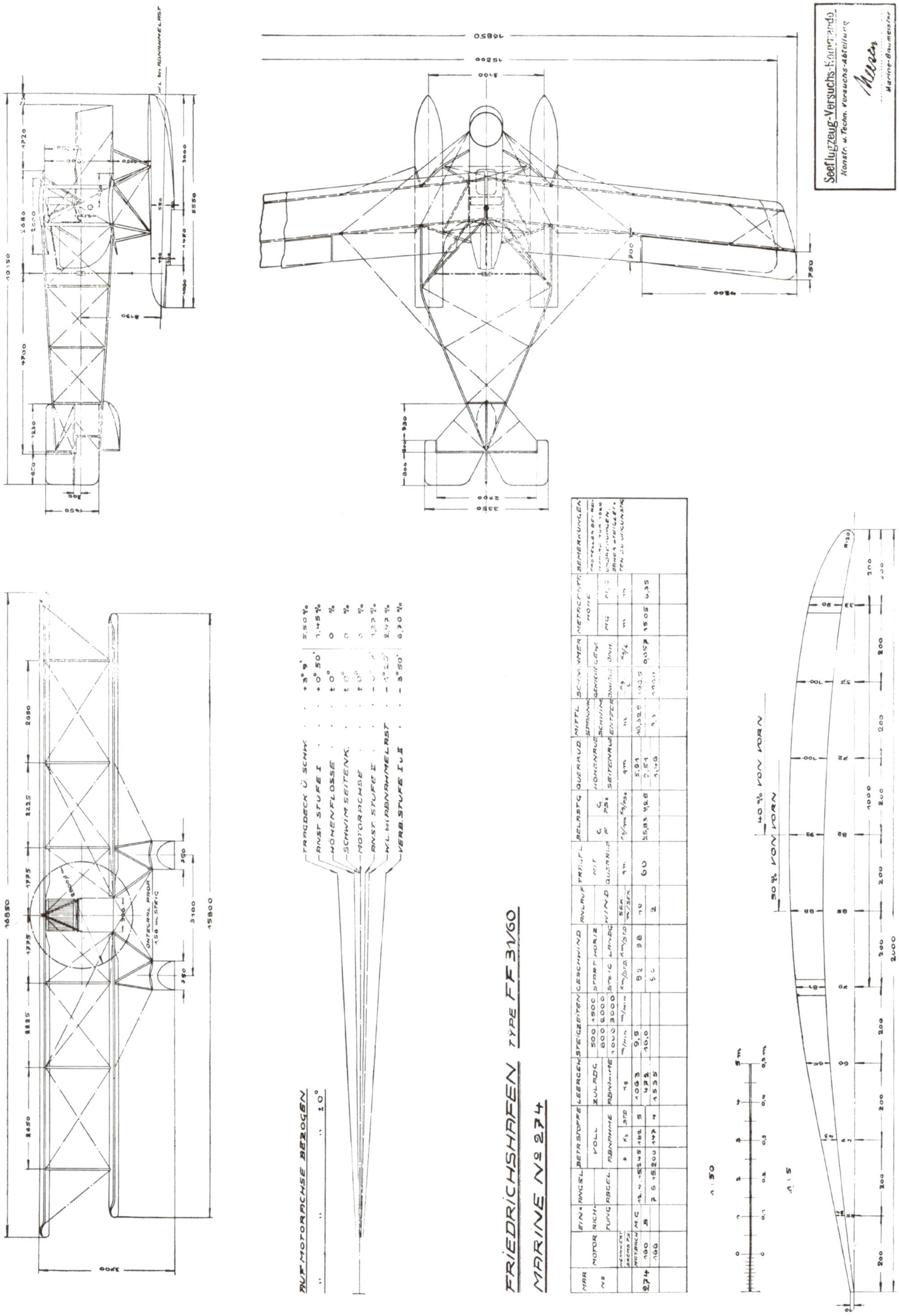
Seeflugzeug-Versuchs-Kommando
Konstr. u. Techn. Versuchs-abteilung
Marine-Baumeister
FRIEDRICHSHAFEN TYPE FF 31V60
MARINE No 274
AUF MOTORACHSE BEZOGEN
TRAGDECK Ü. SCHW.
ANST. STUFE I
HÖHENFLOSSE
SCHWIM SEITENK.
MOTORACHSE
ANST. STUFE II
W.L. bei ABNAHMELAST
VERB. STUFE I u II
INTEGRAL PROP.
1:50
1:5
40% VON VORN
30% VON VORN

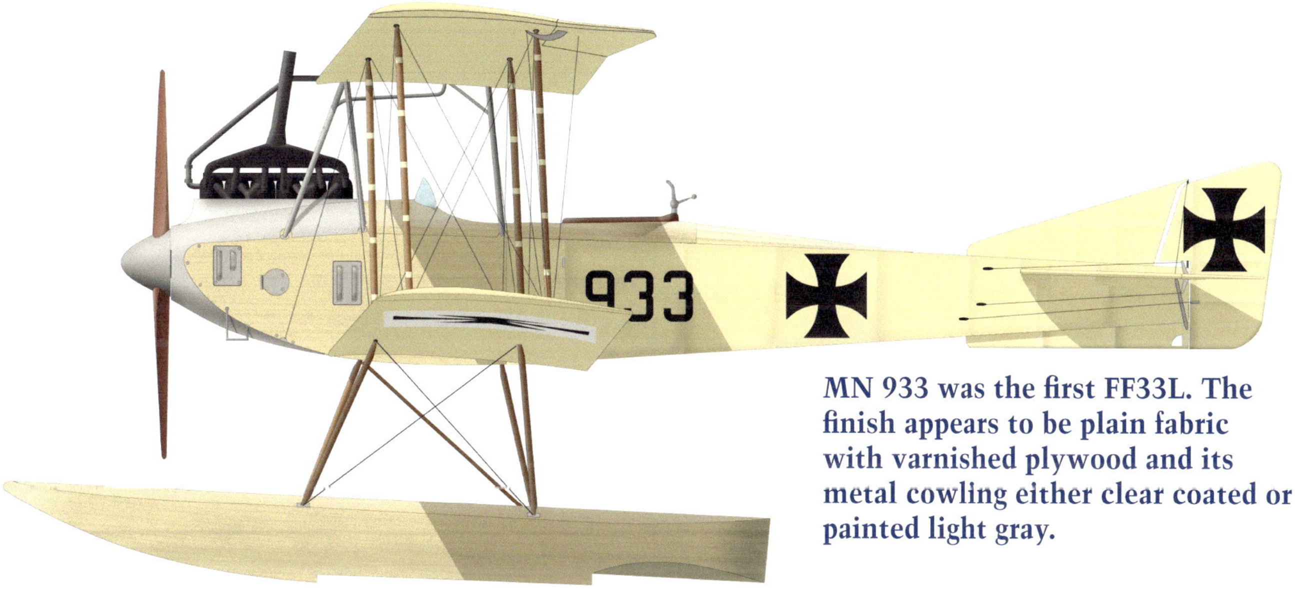

MN 933 was the first FF33L. The finish appears to be plain fabric with varnished plywood and its metal cowling either clear coated or painted light gray.

MN 938 was an FF33L class C2MG based at Zeebrugge. As indicated by his personal markings, it was flown in the Summer of 1917 by Friedrich Christiansen, OC of Zeebrugge, before the arrival of the faster Brandenburg W.12.

MN 1162 was an FF33L class CHFT. It featured a white triangular pennant as a personal or unit marking although the unit and location are unknown. Standard late-war camouflage was applied.

Friedrichshafen FF33 Series

Above: Marine Number 419 was a Friedrichshafen FF33A, while Marine Number 294 was an FF29. The early FF33 and FF33A were virtually FF29 aircraft; later FF33 variants used more powerful engines, mostly the 150 hp Benz Bz.III.

Friedrichshafen's FF33 family was the backbone of German naval aviation during most of the war. Of the 854 seaplanes Friedrichshafen provided to the Navy, 427, or half, were from the FF33 family. Friedrichshafen's seaplanes were consistently more robust and reliable than those from competing manufacturers and had good flying and seakeeping qualities, the key reasons Friedrichshafen aircraft dominated the Navy's production orders. While combat aviators greatly appreciate sparkling performance, reliability is even more important, and that is especially true of seaplanes. Survival time in cold northern waters is often measured in minutes if the seaplane lets the crew down, so reliability is a seaplane's most important quality. Friedrichshafen's seaplanes did not have the exceptional high-altitude performance of Rumpler reconnaissance planes nor the remarkable maneuverability of the Fokker Triplane, but they excelled in reliability. That gained them leadership in the competition to provide reconnaissance seaplanes to the German Navy. And Friedrichshafen's FF33 family built that leadership.

The first orders for Friedrichshafen's FF33 were placed in December 1914, with the first aircraft delivered in April 1915. The final FF33 orders were placed in March 1918, with deliveries completed in November 1918. Given the pace of aircraft development during the war, this longevity of one design was remarkable and a testament to its flying qualities and robust structure.

Reconnaissance & Fighter Variants

The FF33 series included both 3-bay aircraft designed for reconnaissance or training (FF33, FF33B, FF33E, FF33J, and FF33S) and armed two-bay aircraft intended for fighting (FF33F, FF33H, and FF33L). It is still uncertain why all these designs were called variants of the basic FF33 and what the common design elements were throughout the entire series. Certainly Army aircraft types with smaller differences between them were given different type designations.

On the other hand, Army aircraft were generally ordered in large batches compared to Navy aircraft, for which a typical production batch was five or ten aircraft. This resulted in significant variations between production batches of naval aircraft of the same type.

FF33 & FF33A

The initial FF33 order was placed in December 1914 for the variant known as the FF33A. These six aircraft, MN 416–421, were all powered by the 100 hp Mercedes D.I. The FF33 variant, MN 426 and 451–455, were powered by the 120 hp Mercedes D.II. Actually, they were under-powered and subsequent aircraft in the FF33 series had more power.

FF33B & FF33D

The next members of the FF33 family were the FF33B and FF33D. In an endeavor to provide enough power, these aircraft were powered by the 160 hp Maybach. The FF33B were MN 456–458, class BFT, and the FF33D were MN 459–460, class CB; at this time class CB meant an armed two-seat reconnaissance type. At this early stage of the war Maybach engines were primarily intended for airships and production capacity was insufficient to support naval aircraft requirements in addition; no more FF33 aircraft received Maybach engines.

FF33C

The next production order for FF33 series aircraft was for the FF33C, MN 471–473. These aircraft were class BFT and received 160 hp Mercedes D.III engines, the first FF33 series aircraft to receive the more powerful Mercedes and also the last until the Roland-built FF33S trainers were ordered in November 1917. Despite their Mercedes engines, these aircraft were the production prototypes for the FF33E aircraft that was powered by the 150 hp Benz Bz.III. In fact, the official *Atlas deutscher und ausländer Seeflugzeuge* designated MN 471–473 as FF33E aircraft (see the *SVK* table on page 6), but *SVK* war diary records them as being delivered as FF33C aircraft. MN 471 was delivered to the *SVK* on April 6, 1915 and was accepted on April 10 pending successful sea-keeping based on the tests to come. Subsequently, FF33C MN 472 arrived at *SVK* on May 11 and was accepted on May 18, and FF33C MN 473 was delivered on May 22 and accepted on June 19.

The three FF33C aircraft were intensively tested by the *SVK* and did so well all three were sent to operational units. MN471 went to Borkum, MN472 went to Zeebrugge, and MN473 flew to Putzig.

FF33E

With the success of the FF33C aircraft during testing at the *SVK*, the first FF33E production order soon followed. The main difference between the types was the 150 hp Benz used in the FF33E aircraft, further Mercedes engines being assigned to Army aircraft until late in the war. MN 479 was the first FF33E to reach operational service, arriving at Libau naval air station on August 18, 1915.

The FF33E was ordered to fulfill a number of duties. Those class B could carry a few light bombs and were used for operations and training. Those class BFT could carry light bombs and a wireless transmitter, and were used for armed

German Naval Aircraft Classes

Class	Meaning
B	Two-seat aircraft with bombing equipment
BFT	Two-seat aircraft with bombing equipment and wireless transmitter
Bu	U-boat aircraft (*U-Boot Flugzeug*)
C	Two-seat aircraft with one flexible machine gun
C2MG	Two-seat aircraft with one flexible and one fixed machine gun
C3MG	Two-seat aircraft with one flexible and two fixed machine guns
CHFT	Two-seat aircraft with one flexible machine gun and wireless transmitter and receiver
CK	Two-seat aircraft with one flexible cannon and one or two fixed machine guns
E	Single-seat flying boat (*Einsitzerboot*)
ED	Single-seat aircraft with twin floats (*Einsitzer mit Doppelschwimmer*)
ED2MG	Single-seat aircraft with twin floats and two machine guns
FL	Aircraft for control of wire-guided attack boats (*Fernienkflugzeug*)
G	Twin-engine aircraft (*Grossflugzeug*)
HFT	Two-seat aircraft with bombing equipment and wireless transmitter and receiver.
R	Aircraft with three or more engines (*Riesenflugzeug*)
S	Two-seat training aircraft with dual controls (*Schulflugzeug*)
T	Torpedo aircraft (*Torpedoflugzeug*)
U	Practice aircraft (*Uebungsflugzeug für Alleinflieger*)
V	Test aircraft (*Versuchsflugzeug zur Erprobung von Motoren, Propellern, Instrumente, usw.*)

Friedrichshafen FF33 Series Production				
Type	**Marine Numbers**	**Quantity**	**Date Ordered**	**Naval Class & Notes**
FF33A	416–420	5	December 1914	100 hp Mercedes D.I
FF33A	421	1	December 1914	B, 100 hp Mercedes D.I
FF33	426	1	November 1914	B, 120 hp Mercedes D.II
FF33	451–455	5	January 1915	B, 120 hp Mercedes D.II
FF33B	456–460	5	January 1915	CB, MN 459–460 were the FF33D
FF33C	471–473	3	March 1915	BFT, 160 hp Mercedes D.III
FF33E	478–483	6	June 1915	BFT
FF33E	491–496	6	June 1915	BFT
FF33E	497–508	12	August 1915	BFT
FF33E	510	1	August 1915	C
FF33E	529–533	5	October 1915	B
FF33F	534–538	5	October 1915	C
FF33E	541–550	10	November 1915	BFT
FF33E	553–562	10	November 1915	BFT
FF33E	591–595	5	January 1916	BFT
FF33H	596–600	5	January 1916	C
FF33E	633–637	5	January 1916	BFT
FF33H	638–642	5	January 1916	C
FF33E	659–663	5	March 1916	BFT
FF33H	664–668	5	March 1916	C
FF33E	680–684	5	April 1916	BFT
FF33H	685–689	5	April 1916	C
FF33E	690–694	5	April 1916	BFT
FF33H	695–699	5	April 1916	C
FF33E	715–724	10	May 1916	BFT
FF33E	725–729	5	May 1916	HFT
FF33H	730–734	5	May 1916	C
FF33E	736–740	5	May 1916	HFT
FF33H	741–745	5	May 1916	C
FF33S	789–790	2	November 1917	S, Roland built, Mercedes D.III
FF33E	802–811	10	May 1916	HFT
FF33H	812–821	10	May 1916	C
FF33E	822–841	20	June 1916	HFT
FF33E	850–859	10	July 1916	HFT
FF33E	860–869	10	July 1916	B
FF33E	870–881	12	August 1916	HFT
FF33E	882–884	3	September 1916	HFT
FF33E	885–889	5	August 1916	B
FF33J	922–929	8	September 1916	HFT
FF33J	930–931	2	September 1916	HFT

Friedrichshafen FF33 Series Production				
Type	Marine Numbers	Quantity	Date Ordered	Naval Class & Notes
FF33L	932–941	10	September 1916	C2MG; 932 was the sole FF33K
FF33L	1001–1010	10	November 1916	CHFT
FF33L	1085–1094	10	October 1916	CHFT
FF33J	1095–1104	10	October 1916	HFT
FF33L	1117–1126	10	December 1916	C2MG
FF33L	1158–1177	20	January 1917	CHFT, revised tail
FF33L	1234–1278	45	February 1917	CHFT
FF33L	1279–1288	10	February 1917	C2MG
FF33J	1289–1298	10	February 1917	HFT
FF33E	1577–1596	20	June 1917	C2MG
FF33E	1980–1999	20	September 1917	S
FF33S	3001–3030	30	December 1917	S
FF33J	3031–3060	30	March 1918	HFT
FF33L	3144–3153	10	Late 1918	C2MG
FF33L	3154–3193	40	Late 1918	C2MG; order cancelled December 1918
FF33S	6501–6520	20	November 1917	S, Roland built, Mercedes D.III
Note: Unless noted otherwise in the tables, all FF33 series aircraft had the 150 hp Benz Bz.III. Some early FF33 aircraft had Mercedes or Maybach engines prior to an agreement between the Army & Navy that reserved Mercedes engines for Army aircraft and allocated some Benz engines to the Navy. Later, when Mercedes production caught up to demand somewhat, a few naval aircraft like the FF33S aircraft built by Roland were given Mercedes engines.				

reconnaissance. Those class HFT could carry light bombs and a wireless transmitter and receiver, and like their class B siblings were used for armed reconnaissance. Only one FF33E, MN 510, was class C, an armed two-seater.

FF33F
Essentially, the FF33F was the armed version of the FF33E; the FF33F was class C, an armed two-seater with one flexible gun. Only one small batch of FF33F aircraft, MN 534–538, were built.

FF33H
The next member of the FF33 series was the FF33H, the FF33G designation apparently not being used for an aircraft actually built. The FF33H aircraft were class C, armed two-seaters with one flexible gun. However, they differed from previous variants in that they were more compact, two-bay aircraft instead of three-bay aircraft. The intention of this design change was to improve their maneuverability and speed during combat, for which they were armed, but without a fixed gun for the pilot. The FF33H was built in a series of small batches interspersed with batches of the FF33E; see the tables for details.

FF33J
The FF33J was the next production variant of the series. It was a three-bay aircraft, class HFT, and designed for the same roles as the FF33E. Small batches (MN 922–931, MN 1095–1104, MN 1289–1298, and MN 3031–3060) totalling 60 aircraft were built. The FF33J had a more streamlined nose than the FF33E with a propeller spinner the FF33E lacked and also had a streamlined headrest for the observer.

FF33K
The FF33K was a step toward the FF33L. It was a two-bay, class C2MG aircraft. Only one aircraft, Marine Number 932, was built as the first aircraft of the first FF33L production batch. The FF33K had a larger wingspan than the definitive FF33L and its tail surfaces were similar to the older FF33 series aircraft instead of the more refined surfaces of the FF33L.

FF33L
The final and best FF33 variant was the FF33L, a two-bay floatplane with a more streamlined nose for better performance. The first order was placed in September 1916, and ordered continued into February 1917. A total of 125 FF33L aircraft were

Friedrichshafen FF33 Subtype Specifications				
Specification	FF33	FF33A	FF33B & D	FF33E & F
Engine	120 hp Mercedes D.II	100 hp Mercedes D.I	160 hp Maybach Mb.III	150 hp Benz Bz.III
Span, Upper	16.75 m	16.75 m	16.80 m	16.80 m
Span, Lower	—	—	—	15.15 m
Wing Area	—	—	60.00 m^2	52.70 m^2
Length	—	—	10.20 m	10.27 m
Height	—	—	3.73 m	3.73 m
Empty Weight	1,000 kg	1,000 kg	1,010 kg	984–1,010 kg
Loaded Weight	1,600 kg	1,600 kg	1,655 kg	1,635–1,675 kg
Max Speed	110 km/h	—	100 km/h	119–128 km/h
Cruise Speed	—	—	—	110 km/h
Range	—	—	—	600–700 km
Climb to 500 m	—	—	—	5–7 minutes
Climb to 1,000 m	—	—	—	9.9–15.5 minutes
Climb to 1,500 m	—	—	—	16–28 minutes
Note: FF33E/F weight and performance specifications vary depending on production batches.				

built, 40 in class C2MG and 85 class CHFT. The FF33L was fairly agile for its size and the class C2MG aircraft, with their fixed gun for the pilot, were particularly suited for escort and offensive use. The FF33L did not possess the maneuverability or speed of the later Brandenburg W.12, but the FF33L was more robust and could operate in higher sea states than the W.12.

Like many naval aircraft, the FF33L underwent modifications between its relatively small production batches, the most noticeable of which were changes in the vertical tail surfaces. The first change started with batch 1158–1177, which featured a new tail design that eliminated the fin. FF33L #3144 was also tested with new tail surfaces, although these was not adopted for production.

FF33S

The dual-control FF33S was the dedicated training version of the FF33E. One batch, MN3001–3030 was ordered from Friedrichshafen and two from Roland (789–790 and 6501–6520) for a total of 52 aircraft.

Facing Page: Friedrichshafen FF33B Marine Number 458 was a powered by a 160 hp Maybach Mb.III. Only three FF33B aircraft, MN 456–458, were built; they were class BFT. The FF33 was under-powered with only 100–120 hp, and installation of the 160 hp Maybach was an attempt to give the aircraft adequate power.

Left: Marine Number 453 was a Friedrichshafen FF33 powered by a 120 hp Mercedes D.II.

Friedrichshafen FF33 Subtype Specifications				
Specification	**FF33H**	**FF33J**	**FF33K**	**FF33L**
Engine	150 hp Benz Bz.III	150 hp Benz Bz.III	150 hp Benz Bz.III	150 hp Benz Bz.III
Span, Upper	14.20 m	16.70 m	14.20 m	13.20 m
Span, Lower	12.66 m	15.00 m	12.66 m	11.70 m
Wing Area	43.40 m^2	52.50 m^2	43.40 m^2	40.54 m^2
Length	9.60 m	10.30 m	9.60 m	8.95 m
Height	3.75 m	3.72 m	—	4.00 m
Empty Weight	948 kg	1,045 kg	948 kg	917 kg
Loaded Weight	1,477 kg	1,585 kg	1,477 kg	1,388 kg
Max Speed	123 km/h	125 km/h	135 km/h	139 km/h
Cruise Speed	—	115 km/h	—	—
Range	615 km	550 km	675 km	450 km
Climb to 500 m	4.5 minutes	—	7 minutes	—
Climb to 800 m	—	—	12.5 minutes	6 minutes
Climb to 1,000 m	11 minutes	—	16.5 minutes	8 minutes
Climb to 1,500 m	18.5 minutes	—	—	13.5 minutes

Above: Friedrichshafen FF33D Marine Number 460 was a powered by a 160 hp Maybach Mb.III. Only two FF33D aircraft, MN 459–460, were built; they were class CB, meaning an armed two-seater. Later the class designation for an armed two-seater was shortened to class C. The FF33D was the first FF33 variant to be armed with a machine gun as shown. While assigned to Zeebrügge, on the night of 22/23 Jan. 1916, this aircraft attacked Dover with 10 bombs.

Above: Friedrichshafen FF33D Marine Number 459 ready to launch.

Above: Friedrichshafen FF33D Marine Number 459 in a lineup of FF33 floatplanes outside the factory. The distinctive side radiator for its160 hp Maybach Mb.III is clearly visible. The five FF33B and FF33D floatplanes built were the only FF33 variants powered by the Maybach, which did not have sufficient production capacity at the time to support greater production quantities of aircraft.

Above: Sanke card from a photo of an early Friedrichshafen FF33 variant that bombed Dover.

Top: Friedrichshafen FF33E Marine Number 501 on patrol. This early FF33E retains the tail float featured by earlier designs, but later production batches of the FF33E would eliminate the tail float for reduced weight and drag.

Above: Friedrichshafen FF33E Marine Number 501 on patrol.

Left: Friedrichshafen FF33E without tail float.

Above: Another image of Friedrichshafen FF33E Marine Number 501 on patrol.

Below: Friedrichshafen FF33E Marine Number 549 with ground crew.

Above: A Friedrichshafen FF33E (or FF33F?) operating with a U-boat.

Above: FF33E Marine Number 559 with old insignia.

Above: Friedrichshafen FF33E Marine Number 553 landing.

Below: Friedrichshafen FF33E Marine Number 553 on the water.

Above: Friedrichshafen FF33E Marine Number 557 on the beach at Warnemünde; it was ordered in November 1915.

Below: Friedrichshafen FF33E Marine Number 634 on its beaching at Warnemünde; it was ordered in January 1916.

This Page: Friedrichshafen FF33E Marine Number 661 on its beaching dolly at Warnemünde. This aircraft was part of a batch of five ordered in March 1916. On 8 June 1916, sister aircraft FF33H #639, flown by *Flugmeister* Ponater and *Ltn.z.S.* Boenisch, shot down a French flying boat. The two crewmen were rescued by a U-boat and brought to Zeebrügge. On 20 August #639 caught fire in flight and force-landed near Ostend; the observer was slightly burned.

Above: An early model Friedrichshafen FF33E covered with ice. The early FF33 floatplanes like this one had a tail float like the preceding FF29 aircraft.

Above: Friedrichshafen FF33E Marine Number 560 was a class BFT ordered in November 1915. By this production batch the FF33E had lost the tail float used on early FF33E aircraft.

Above: Friedrichshafen FF33E Marine Number 633 was the first of a batch of five aircraft ordered in January 1916 and was class BFT. This aircraft has the tail surface design characteristic of the main FF33E production aircraft, with the rudder and fin below the fuselage larger than those of Marine Number 560.

Below: An FF33E in flight. Unfortunately, the Marine Number is not legible, but the vertical tail surfaces resemble those of Marine Number 633 above. The FF33E was the iconic mid-war German floatplane.

Above & Below: Friedrichshafen FF33E Marine Number 717 was a class BFT ordered in May 1916 and features the larger vertical tail without tail float.

Above: Friedrichshafen FF33E Marine Number 729 was a class HFT ordered in May 1916 and has the larger vertical tail.

Below: Friedrichshafen FF33E Marine Number 738 operating with a U-boat; it was also a class HFT ordered in May 1916.

Above: A striking view of a Friedrichshafen FF33F. Unfortunately, the Marine Number to verify that is not fully visible, but there appears to be a gun mounting for the observer. The FF33F was the armed version of the FF33E.

Left: The observer of a Friedrichshafen FF33F passes a message to a U-boat. Again there appears to be a gun-ring and gun mounted for the observer.

Right: A Friedrichshafen FF33E in Bulgarian service.

Right: Friedrichshafen FF33E Marine Number 871 was a class HFT ordered in August 1916.

Above & Right: Friedrichshafen FF33E Marine Number 1981 was a class S, or training aircraft with dual controls, ordered in September 1917. It was part of batch 1980–1999. Subsequent FF33 series aircraft ordered as dual-control trainers were given the designation FF33S and were essentially the same as the dual-control FF33E.

Above: Friedrichshafen FF33H Marine Number 596 was the first production FF33H. A class C, it was ordered in January 1916. The FF33H was a more compact design than the FF33E, having two bays of struts compared to the three bays of struts used in previous FF33 variants.

Below: Friedrichshafen FF33H Marine Number 600 was the last aircraft in the first production batch of FF33H floatplanes. A class C, it was ordered in January 1916 along with #596 above. The intention of the reduced span of the FF33H was to improve maneuverability of this armed sub-type to make it more effective during air-to-air combat.

Left: Friedrichshafen FF33H Marine Number 695 flying a mission from Angersee. This aircraft, an armed class C, was the first aircraft of the 5th FF33H production batch ordered in April 1916. The red identification pennants can be seen attached to both #695 and the aircraft from which the photo was taken.

Below: Friedrichshafen FF33H Marine Number 695 at Angersee.

Above: Another view of Friedrichshafen FF33H Marine Number 695.

Below: Friedrichshafen FF33H Marine Number 699 was the last aircraft in the production batch of FF33H floatplanes ordered in April 1916 that included #695 and was also an armed class C.

Above: Friedrichshafen FF33H Marine Number 731 in flight (left) and taxiing (right) was ordered in May 1916.

Left: Friedrichshafen FF33H Marine Number 819 was from another production batch also ordered in May 1916. Like other FF33H aircraft it was an armed class C floatplane.

Above & Below: Two views of Friedrichshafen FF33H Marine Number 744 serving at Angersee. The observer's gun is clearly visible above, and both views illustrate additional markings that were rare for these mostly anonymous floatplanes.

Right: Friedrichshafen FF33H; the Marine Number is not visible but the observer's gun ring is clearly shown.

Right: Posed propaganda photograph of a Friedrichshafen FF33H 'rescuing' a downed airman. There appears to be a cockade on the fuselage of the crashed 'floatplane', which is a mockup.

Below: Front view of a Friedrichshafen FF33H emphasizes its design heritage from earlier FF33 variants despite its two-bay wings.

Above& Below: Friedrichshafen FF33J Marine Number 3033 was a class HFT ordered in March 1918, a remarkable achievement for a basic design that originated in 1914. Unusually for a slow floatplane, the observer has a streamlined headrest. Late-war national insignia are prominent along with the streamlined nose with propeller spinner.

Bottom: FF33J Marine Number 303x, a class HFT aircraft, preparing for a mission.

Left & Above: Two views of Friedrichshafen FF33J Marine Number 930, the first FF33J built, on hoists. A three-bay design to class HFT, the FF33J was basically an FF33E with a more streamlined nose. The streamlining improved cruise speed by 5 km/h. Only four small production batches totalling 30 FF33J aircraft were built. (Above photo Peter M. Bowers Collection/Museum of Flight)

Above: A rear quarter view of Friedrichshafen FF33J Marine Number 1099 being launched using a hoist. The FF33J was an unarmed class HFT and the observer had a streamlined headrest.

This Page: Marine Number 932 was the only Friedrichshafen FF33K built. A class C2MG, it was an intermediate step to the definitive FF33L and was built as the first aircraft of the first FF33L production batch ordered in September 1916. The FF33L was a smaller, lighter, more refined design than the FF33K and offered better performance and maneuverability, so was the variant produced in quantity for the more aggressive C2MG role.

Above & Below: Marine Number 933 was the first Friedrichshafen FF33L built. A class C2MG, it was a smaller, lighter aircraft than the FF33K that was slightly faster and had double the rate of climb. It was also more maneuverable, all qualities that enhanced its effectiveness for its designed role as the Navy's first two-seat floatplane for the C2MG fighting role.

Above & Below: As the first Friedrichshafen FF33L built, Marine Number 933 was the subject of a number of photos. These show the type's more compact, streamlined form and fixed pilot's gun to advantage.

Above: A Friedrichshafen FF33L being launched by crane. If it was being recovered, the engine would be stopped; the fact it is running shows the aircraft was being launched. Along with its two-bay wings and smaller size, the FF33L had a refined tail design for greater agility compared to previous models of the FF33 series. (The Peter M. Bowers Collection/The Museum of Flight)

Above: This head-on view of a Friedrichshafen FF33L shows its small frontal area for a floatplane, contributing to low drag.

Above & Below: Marine Number 1001 was the first Friedrichshafen FF33L of the second production batch. This batch was built as class CHFT. The weight of the wireless transmitter and receiver meant something else had to be eliminated, and that was the pilot's fixed gun, which could not be fitted. Of the 125 FF33L aircraft built, 85 were class CHFT. (Lower photo Peter M. Bowers Collection/Museum of Flight)

Above: This lineup of floatplanes at Windau in the Baltic includes Friedrichshafen FF33L Marine Number 1009 at far right and 1239 second from right; both are class CHFT.

Below: Closeup of Friedrichshafen FF33L Marine Number 1239 from the above image. (The Peter M. Bowers Collection/ The Museum of Flight)

Above: This lineup of *Seeflugstaffel* 1 at Angersee in the Baltic includes Friedrichshafen FF33L Marine Number 1123 (class C2MG) in the center foreground, FF33L #1094 (class CHFT) at far right, FF33L #1236 (class CHFT) second from right, FF33L #1091 (class CHFT) second from left, and FF33L #1175 (class CHFT) at far left. The aircraft in center background is FF33E Marine Number 805 (class HFT), the only aircraft in the photo that is not an FF33L.

Above: Another view of Angersee with Friedrichshafen FF33L Marine Number 1123 (class C2MG) at left and another FF33L (#1094?) in the foreground.

Above: Friedrichshafen FF33L#1162 was a class CHFT; the antenna mechanism can be seen alongside the observer's cockpit. (Peter M. Bowers Collection/Museum of Flight)

Above: Friedrichshafen FF33L Marine Number 1235, class CHFT, photographed on patrol.

Above: Friedrichshafen FF33L Marine Number 1174 (class CHFT) is at far left and FF33J Marine Number 1103 (class HFT) is in the center.

Above: Friedrichshafen FF33L Marine Number 1244, class CHFT, without its spinner and approaching the ramp.

Above: Friedrichshafen FF33L Marine Number 1244, class CHFT, without its spinner tying up.

This Page: FF33L #1257, class CHFT, being recovered. This aircraft had the modified tail without fixed fin. From the clothing worn by the civilian onlookers the location was Turkey.

Above: FF33L #1262, class CHFT, with standard vertical tail surfaces. (The Peter M. Bowers Collection/The Museum of Flight)

Below: FF33L #1263, class CHFT, at left foreground and another FF33L class CHFT at right; both aircraft are in the standard late-war German naval aircraft camouflage. Albatros W4 floatplane fighters line the background.

Above: FF33L #1590, class C2MG, had the modified tail without fixed fin.

Below: FF33L #3144, class C2MG, had an experimental enlarged rudder; this photo shows it without a spinner. (The Peter M. Bowers Collection/The Museum of Flight)

Above: FF33L with the markings of Friedrich Christiansen on the crane at Zeebrugge. Christiansen flew FF33L Marine Number 938 frequently until October, 1917, when he switched to the faster, more maneuverable Brandenburg W.12. Although the Marine Number on this aircraft is somewhat obscured, it is thought to be #938, a class C2MG from the first FF33L production batch delivered in February 1917. Christiansen's personal markings were applied to his aircraft after he became Commanding Officer of Zeebrugge. (Peter M. Bowers Collection/Museum of Flight)

Below: Lineup of FF33L floatplanes at Zeebrugge.

Above: FF33L floatplanes from Zeebrugge intercept a Dutch schooner.

Below: FF33L from Zeebrugge inspecting the Dutch schooner.

Above: FF33L V3 in postwar Netherlands service with non-standard vertical tail surfaces.

Below: FF33L V3 in postwar Netherlands service taking off.

Above: FF33L in German naval camouflage fabric; details of thi situation are not known.

Below: FF33L in German markings and camouflage. The designers minimized frontal area for minimum drag.

Above: FF33L V15 in postwar flight over the Netherlands displays the later Dutch national markings.

Below: FF33L V.22 in postwar Netherlands service with the original Dutch national insignia. A Macchi is at the right.

Above: FF33L V23 in postwar Netherlands service carrying the original national markings of an orange disk on the rudder.

Left: FF33L; dashed lines on the tail show were the national insignia will be added.

Below: FF33L in postwar Netherlands service.

Operations at Zeebrugge

At Zeebrugge the floatplanes were stored on land for maintenance access and some protection from the weather. For missions the floatplanes were then transported by train from their storage location to the end of the Mole, where they were launched into the water for take-off. Use of the train enabled many aircraft to be launched quickly, and the locomotive was kept under steam at all times in case of emergency. The images of this train on this page are a series of video captures from a movie of the process. FF33L floatplanes Marine Numbers 1162, class CHFT, and 1281, class C2MG can be identified.

Once the floatplanes reached the end of the Mole, the engines were started by the ground crew and the flight crew boarded the aircraft as shown on this page and the following page. The engines were started manually by the ground crew by swinging the propeller because the aircraft did not have starters. When afloat the observer could start the engine by swinging the propeller as he stood on one of the floats. However, it was much safer and more convenient for the ground crew to swing the propeller when the aircraft was on land.

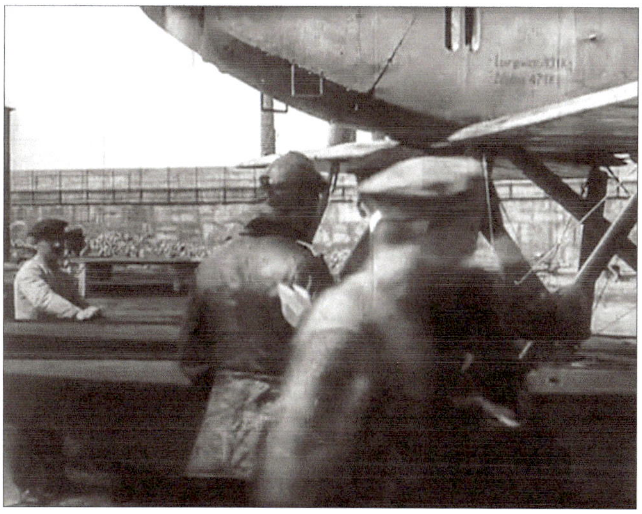

The photos on this page show the completion of crew boarding. Once the crew was in the aircraft, the ground crew handed up a cage of carrier pigeons (*Brieftauben*) to the observer. In these early days of radio, when radio reliability and performance were poor and indeed many aircraft had no radio, carrier pigeons were an essential backup method for long-range communication.

Once the crew had boarded the aircraft with engine already running and the observer had the carrier pigeons and other gear, the floatplane was lifed by crane and placed in the water. As shown on this page, the observer handled the hook, disengaging it after the plane was on the water. The aircraft then taxiied away from the Mole for take-off. This FF33L has striped camouflage on the wings.

The sequence on this page shows the FF33L taxiing away from the Mole for take-off, then lining up into the wind and taking off. The images on the preceding page are too dark to show details of the markings or Marine Number except for diagonal fuselage stripes. The upper photos on this page appear to show the markings of Christiansen's FF33L Marine Number 938.

Once in flight the crew climbed to their patrol altitude while flying to their patrol area. Upon arrival in the patrol area the operational part of the mission started. This sequence shows the observer performing some of his most important functions, visual reconnaissance and in-flight photography.

For filming purposes the aircraft was fastened in place so it could be filmed with the engine running. Interestingly, this aircraft has been fitted with a semi-circular radiator as used in Rumplers instead of the standard rectangular radiator normally fitted. Was this for extra cooling during filming?

The observer had other duties in addition to photography, including navigation and communication if the aircraft was equipped with a radio. Furthermore, as shown on this page, bomb dropping (upper left), attacking targets with the machine gun (upper right), and communicating via carrier pigeon (other photos) were also part of the observer's repertoire.

Another duty for the patrolling FF33L floatplanes was intercepting and engaging enemy aircraft. In these frames from a propaganda film the FF33L (at left) is chasing a Rumpler 6B1 (at right) with cockades painted over its German crosses to represent an Allied aircraft.

The images on this page show FF33L aircraft on patrol, then intercepting a ship.

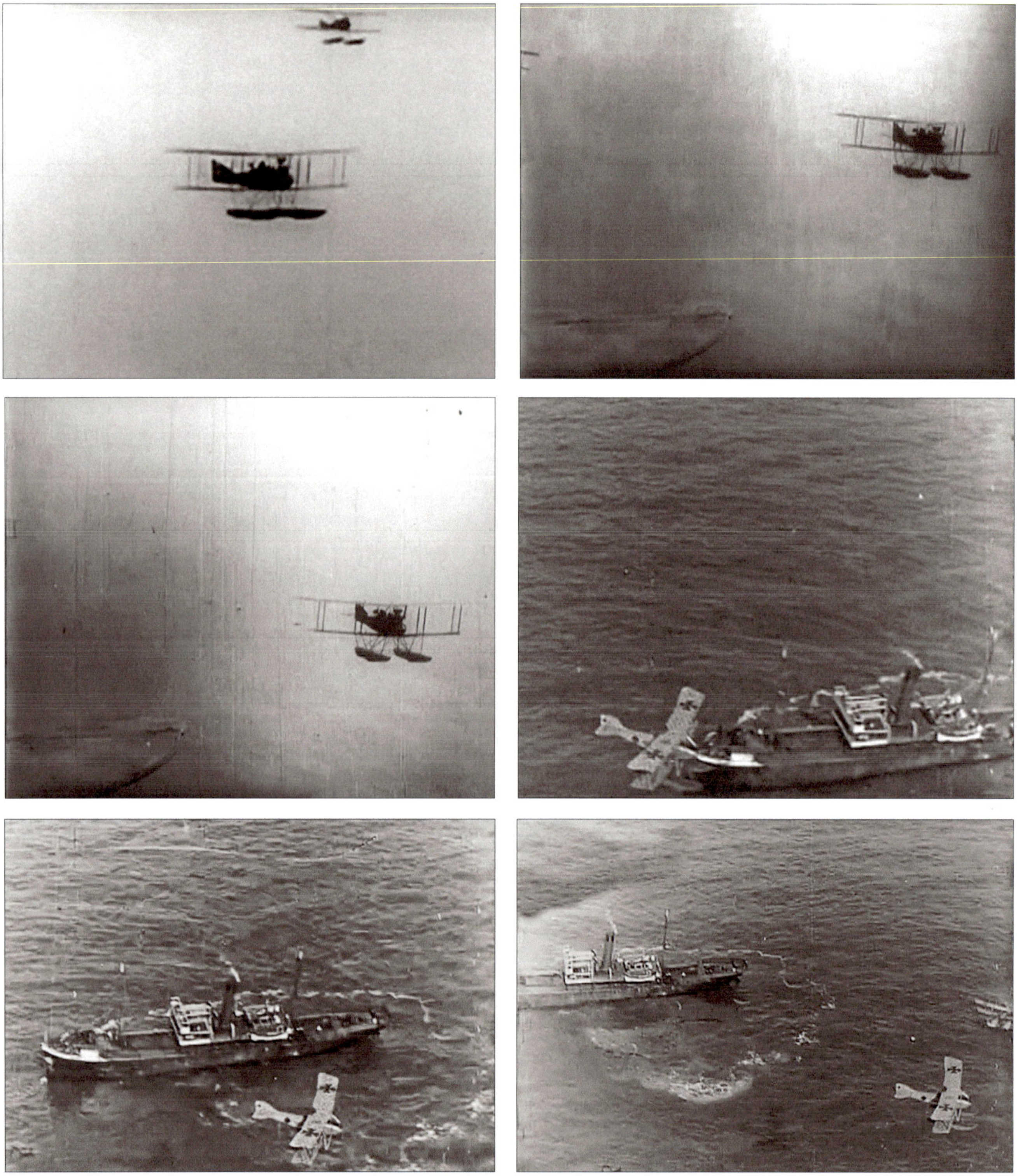

The images on this page record the FF33L floatplanes landing near the merchant vessel and stopping her to inspect her cargo for contraband.

These images, taken from another FF33L floatplane on the surface, show FF33L Marine Number 1118 maneuvering to enable the observer to board the ship for inspection. The ship has heaved-to and lowered a small boat to help the observer board the ship.

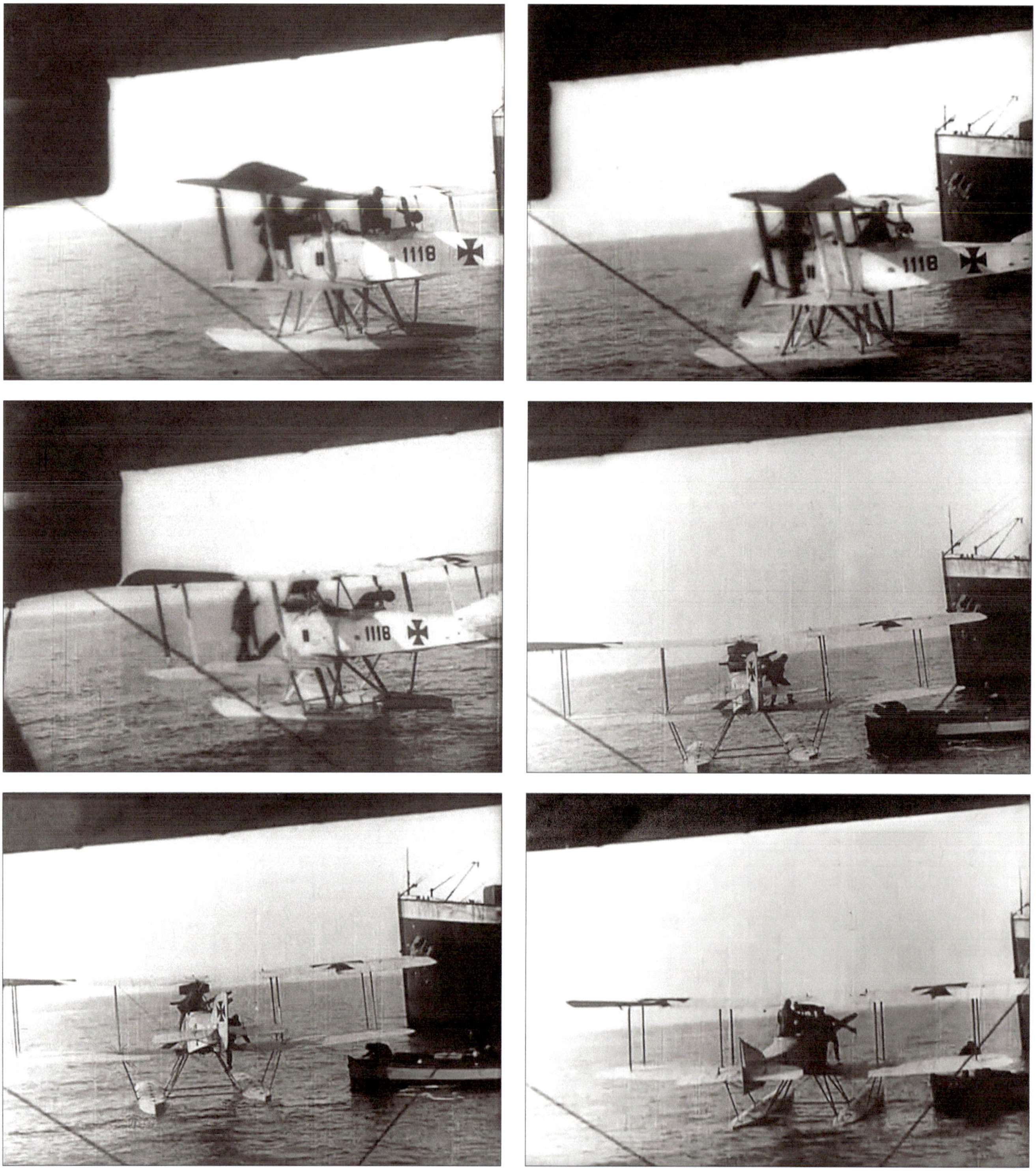

These are more images showing FF33L Marine Number 1118 maneuvering to enable the observer to board the ship for inspection. The ship has heaved-to and lowered a small boat to help the observer board. This was hazardous, especially during heavy seas.

More images showing FF33L Marine Number 1118 maneuvering to enable the observer to board the ship for inspection. One can imagine how hazardous this process was during heavy seas.

These aircraft operated as flights to provide mutual protection during boarding operations and to rescue comrades in event of being forced down at sea due to combat or mechanical problems.

To save fuel and thereby extend the duration of their patrol, the German floatplanes would often land and wait on the water until another patrolling floatplane found a target and returned to summon them. The observers would then need to 'prop' the engine to start it while standing on a float. Shown on this page are frames from a video of an FF33L waiting; note the propeller is not rotating.

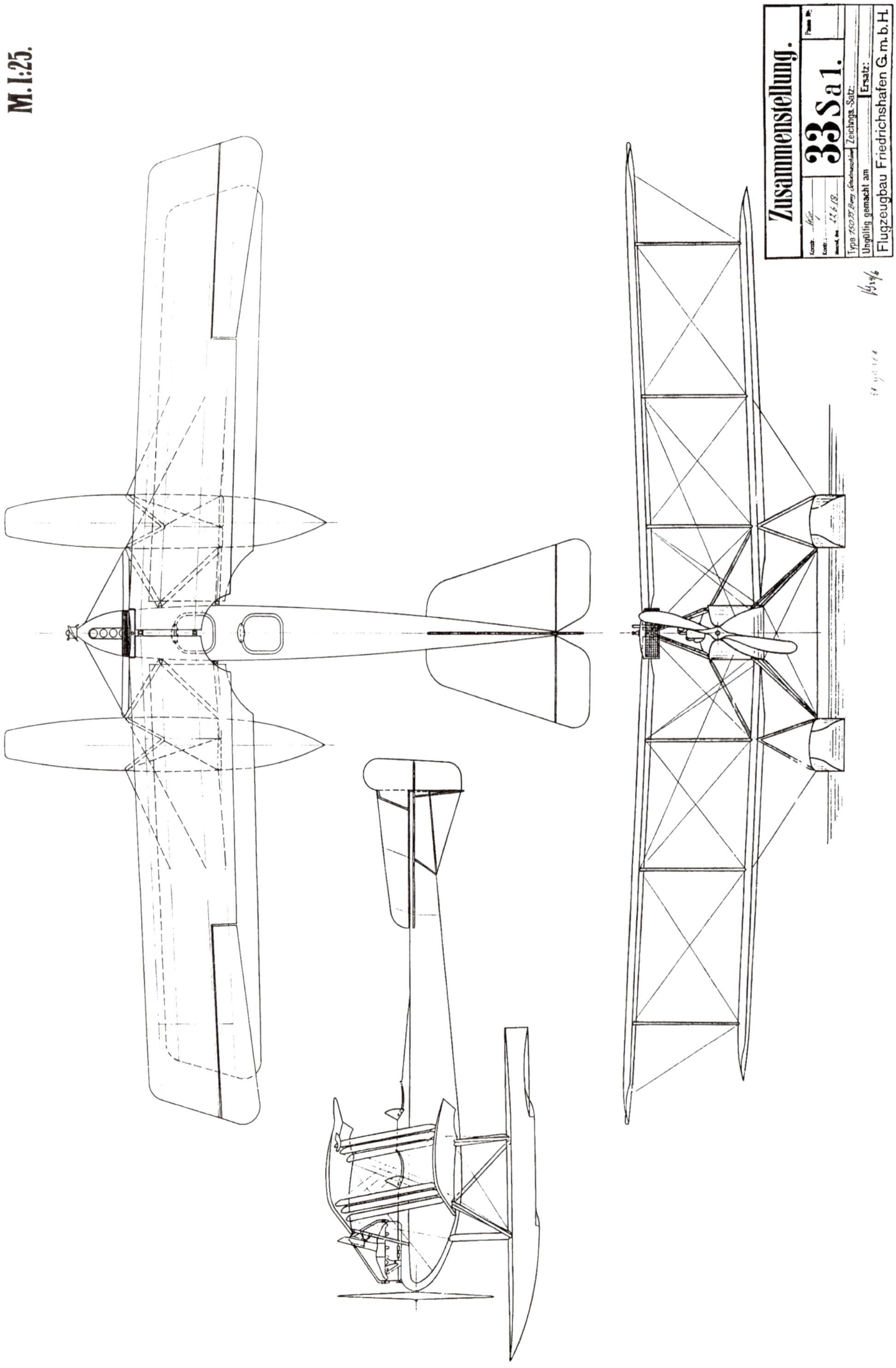
M. 1:25.
Zusammenstellung.
33 Sa 1.
Zeichnungs-Satz:
Ungültig gemacht am
Ersatz:
Flugzeugbau Friedrichshafen G. m. b. H.

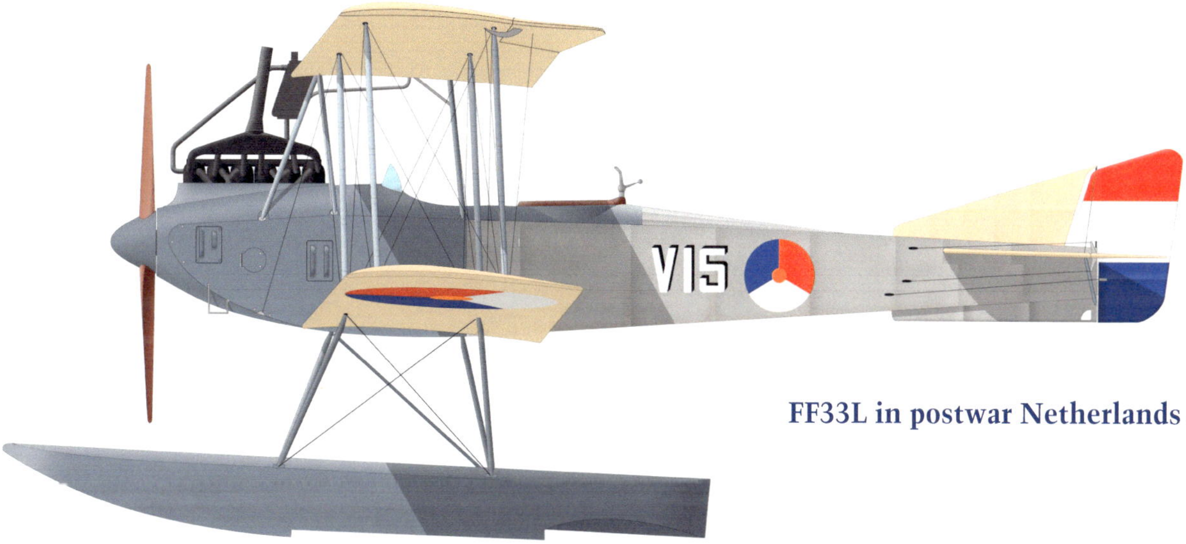

FF33L in postwar Netherlands service.

Right: FF33L in postwar Netherlands service.

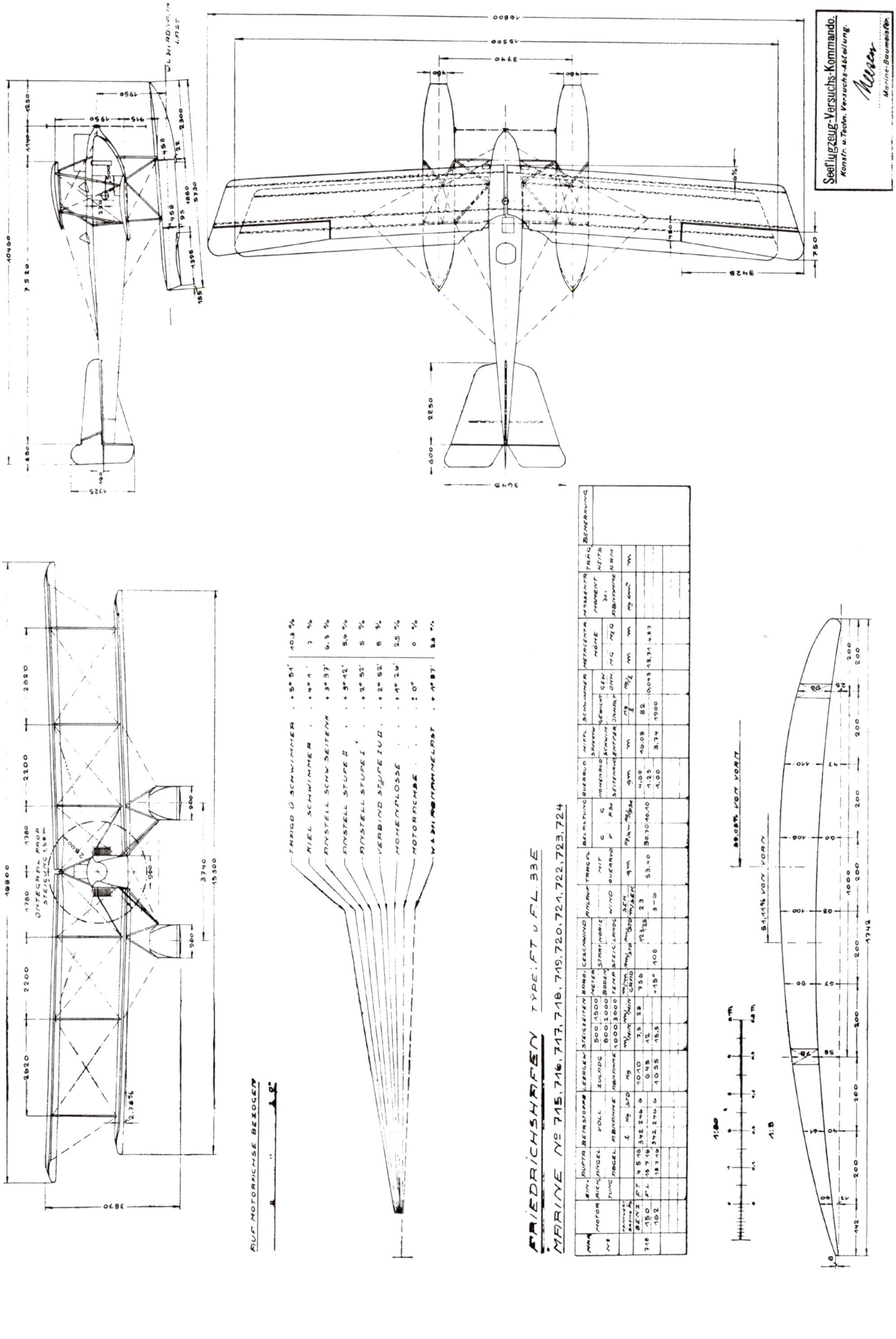
Seeflugzeug-Versuchs-Kommando.
Konstr. u. Techn. Versuchs-Abteilung.
Marine-Baumeister.
FRIEDRICHSHAFEN TYPE: FT u FL 33E
MARINE No 715, 716, 717, 718, 719, 720, 721, 722, 723, 724

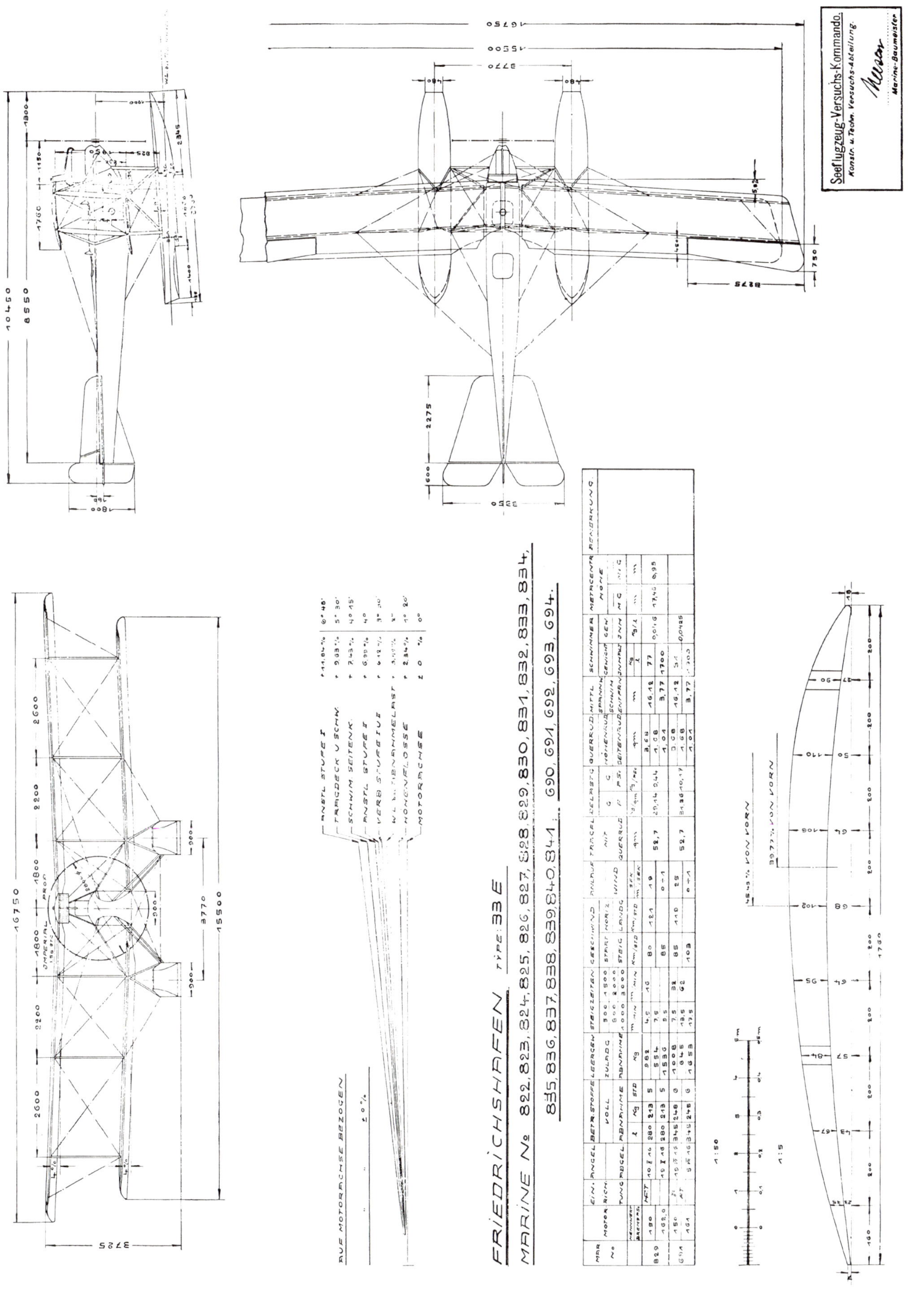
Seeflugzeug-Versuchs-Kommando.
Konstr. u. Techn. Versuchs-Abteilung.
Marine-Baumeister
FRIEDRICHSHAFEN TYPE: 33 E
MARINE No 822, 823, 824, 825, 826, 827, 828, 829, 830, 831, 832, 833, 834,
835, 836, 837, 838, 839, 840, 841, 690, 691, 692, 693, 694.
AUF MOTORACHSE BEZOGEN
16750
15500
3770
3725

Friedrichshafen FF33E SVK Drawing

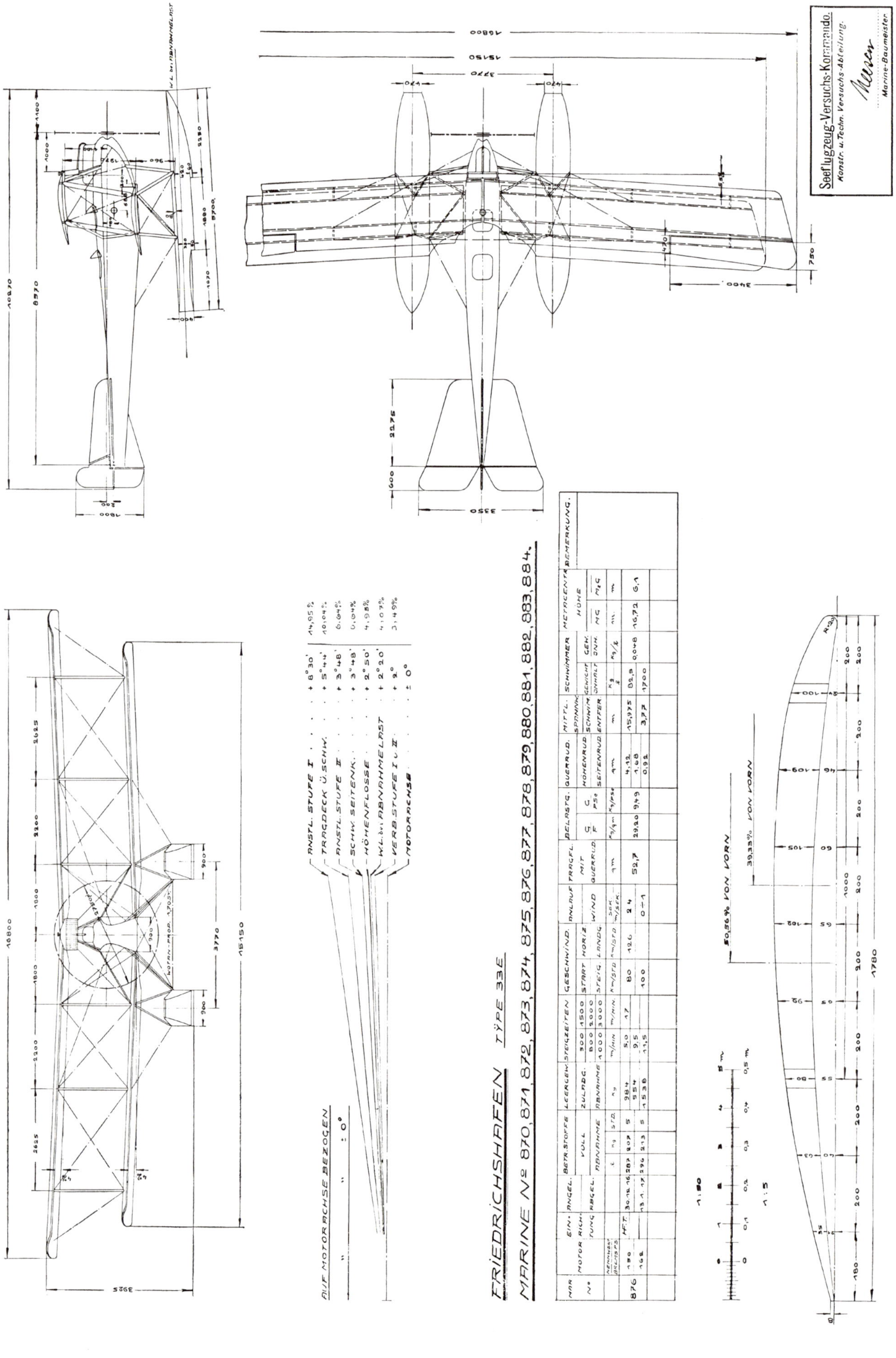

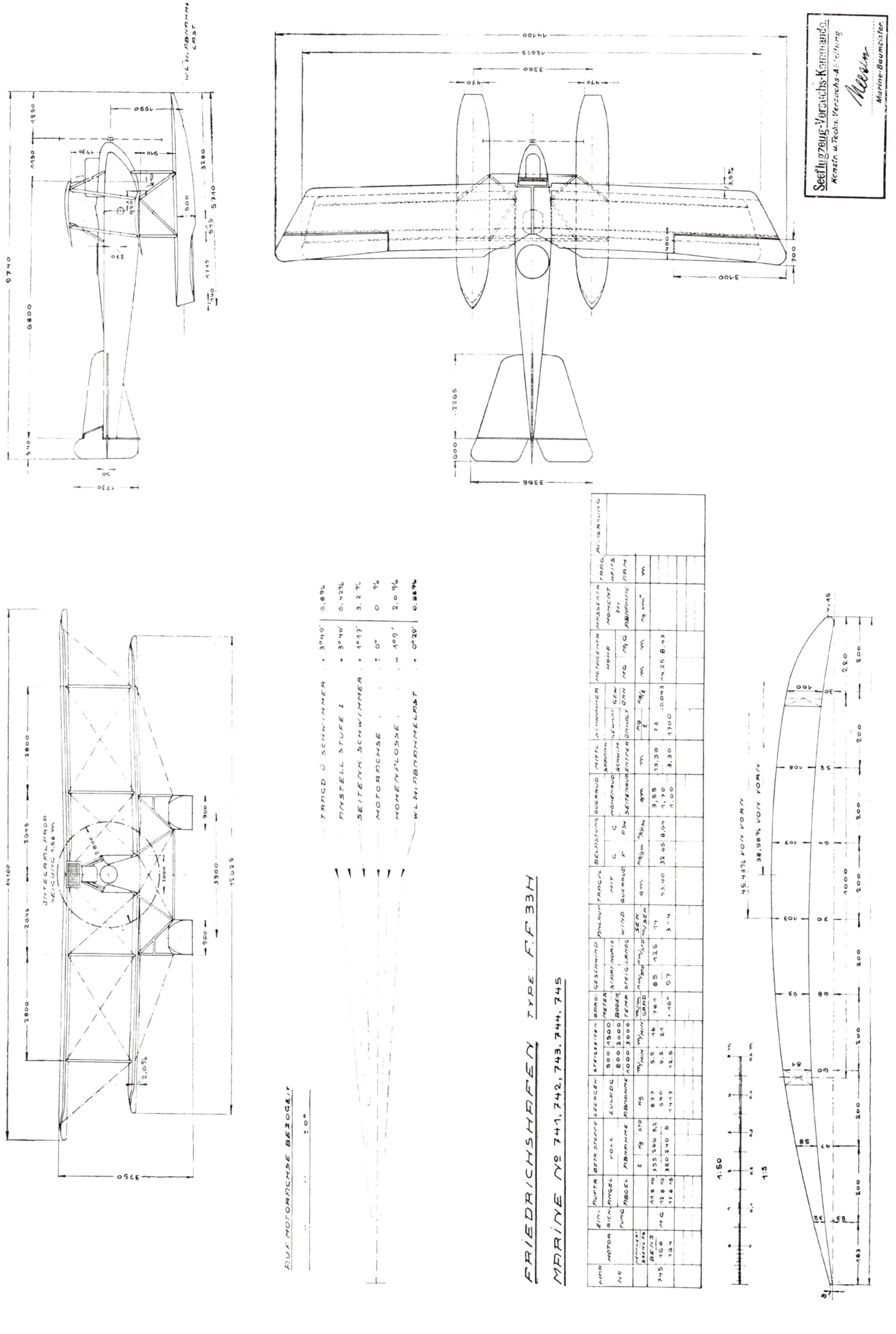
Seeflugzeug-Versuchs-Kommando
Marine-Baumeister
FRIEDRICHSHAFEN TYPE: F.F 33H
MARINE No 741, 742, 743, 744, 745

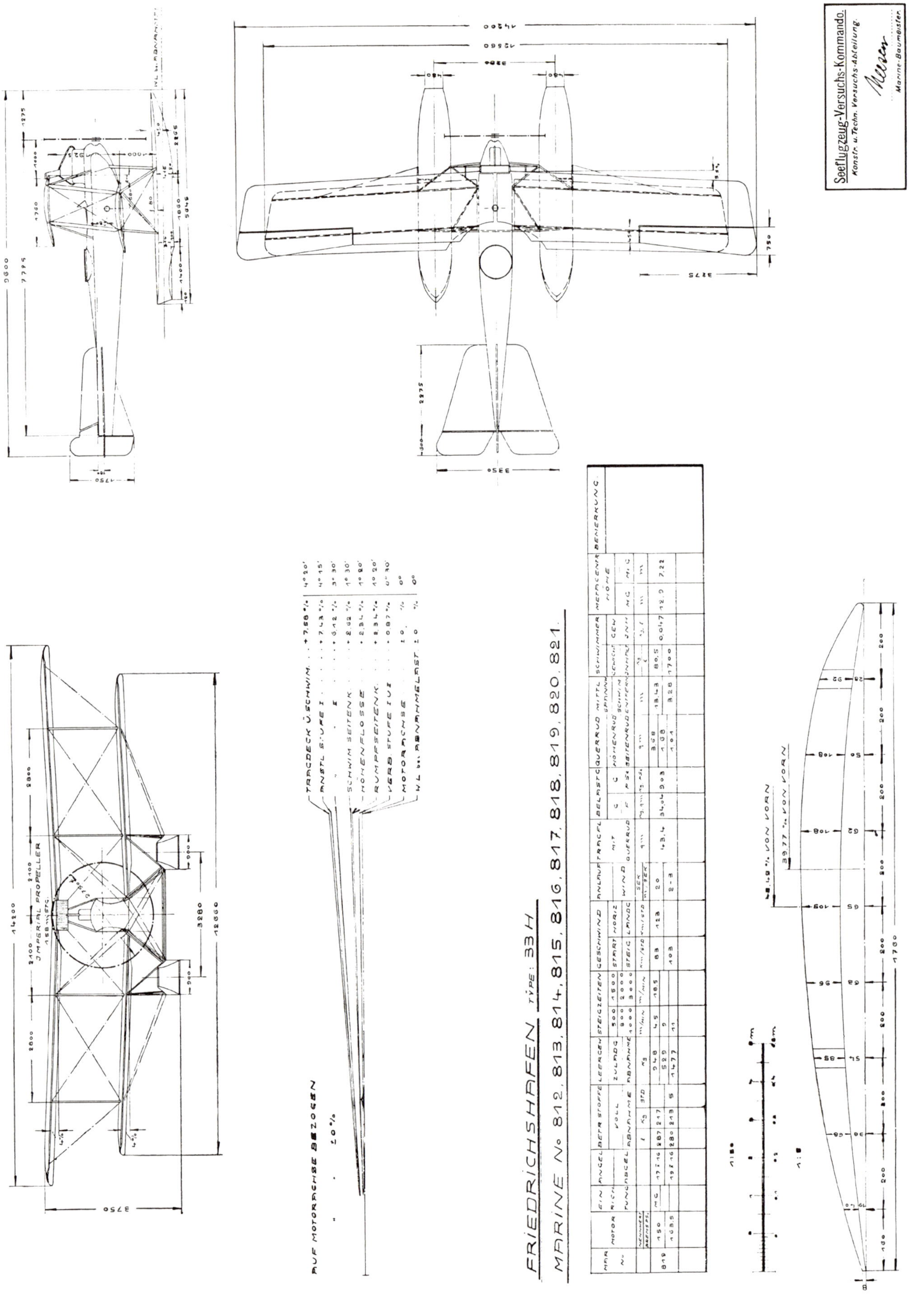
Seeflugzeug-Versuchs-Kommando.
Konstr. u. Techn. Versuchs-Abteilung.
Marine-Baumeister.
FRIEDRICHSHAFEN TYPE: 33 H
MARINE No 812, 813, 814, 815, 816, 817, 818, 819, 820, 821
IMPERIAL PROPELLER

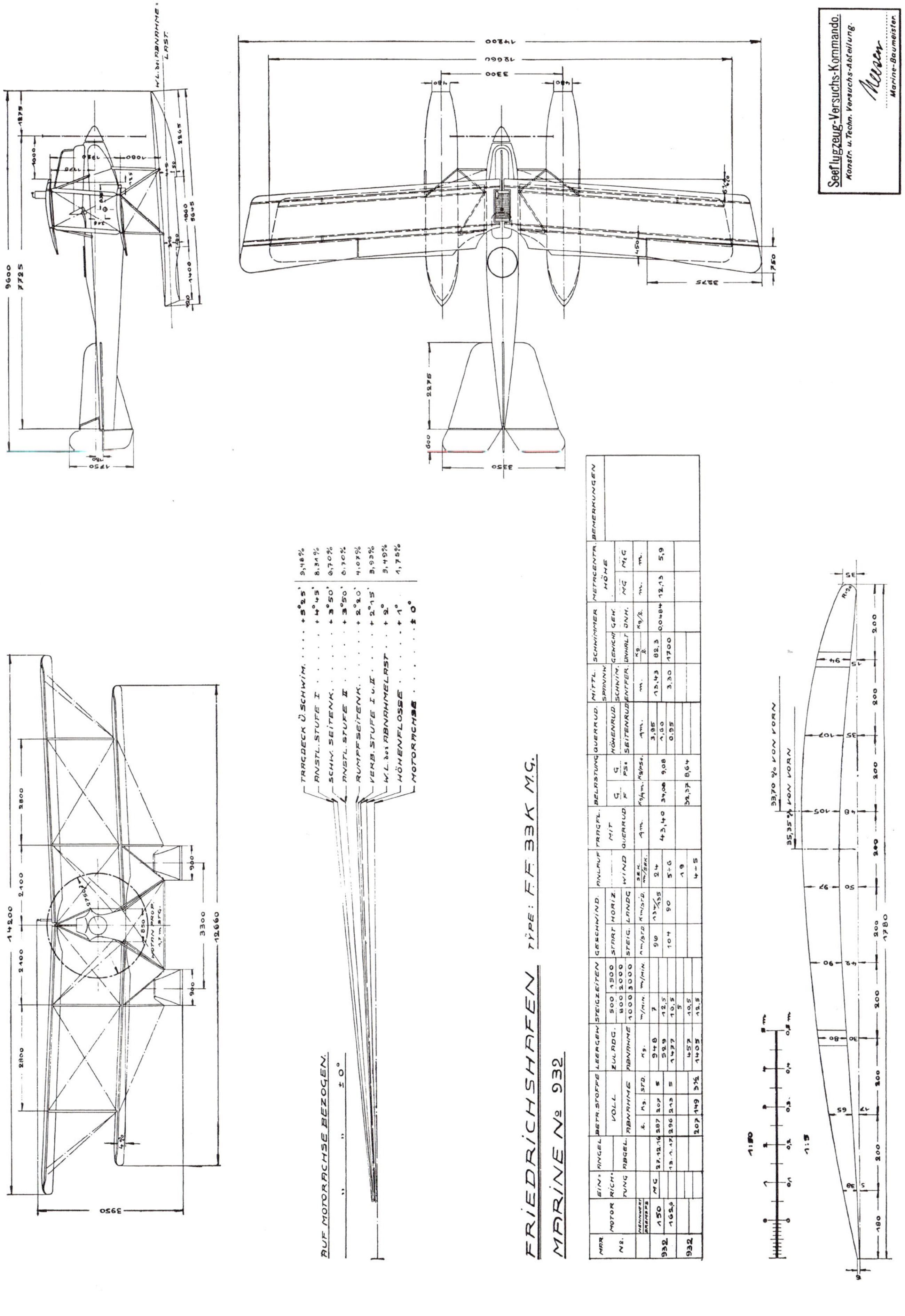
Seeflugzeug-Versuchs-Kommando.
Konstr. u. Techn. Versuchs-Abteilung.
Marine-Baumeister.
FRIEDRICHSHAFEN TYPE: F.F. 33 K M.G.
MARINE No 932
AUF MOTORACHSE BEZOGEN.
TRAGDECK Ü. SCHWIM. +5°25' 9,48%
ANSTL. STUFE I +4°45' 8,31%
SCHW. SEITENK. +3°50' 6,70%
ANSTL. STUFE II +3°50' 6,70%
RUMPFSEITENK. +2°20' 4,07%
VERB. STUFE I u. II +2°15' 3,93%
W.L. bei ABNAHMELAST +2° 3,49%
HÖHENFLOSSE +1° 1,75%
MOTORACHSE ±0°
33,70 % VON VORN
35,35 % VON VORN

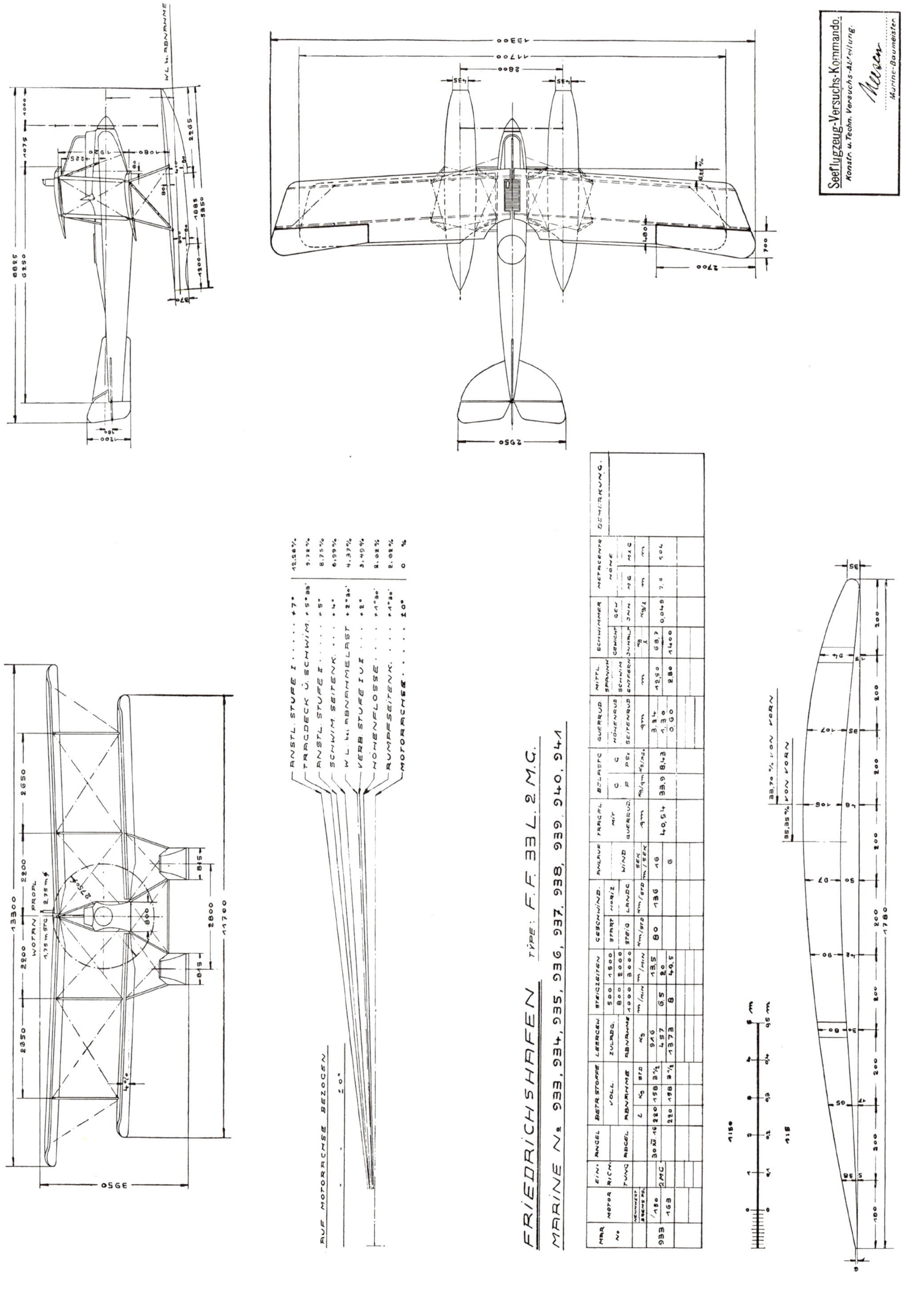
Seeflugzeug-Versuchs-Kommando.
Konstr. u. Techn. Versuchs-Abteilung.
Marine-Baumeister.
FRIEDRICHSHAFEN TYPE: F.F. 33 L. 2 M.G.
MARINE No 933, 934, 935, 936, 937, 938, 939, 940, 941
W.L. bei ABNAHME
WOTAN PROPL. 2,75 m Ø
1,75 m STG.
ANSTL. STUFE I +7° 12,28%
TRAGDECK Ü. SCHWIM. +5°33' 9,72%
ANSTL. STUFE II +5° 8,75%
SCHWIM. SEITENK. +4° 6,99%
W.L. bei ABNAHMELAST +2°30' 4,37%
VERB. STUFE I u. II +2° 3,49%
HÖHENFLOSSE +1°30' 2,62%
RUMPFSEITENK. +1°30' 2,62%
MOTORACHSE ±0° 0%
AUF MOTORACHSE BEZOGEN
MAR. No | MOTOR | EINRICHTUNG | ANGEL. | BETR. STOFFE | LEERGEW. ZULADG. ABNAHME | STEIGZEITEN | GESCHWIND. | ANLAUF | TRAGFL. MIT QUERRUD. | BELASTG. | QUERRUD HÖHENRUD SEITENRUD | MITTL. SPANNW. SCHWIM. ENTFERN. | SCHWIMMER GEWICHT INHALT | METACENTR. HÖHE | BEMERKUNG.
933 | 1/180 163 | 2 MG | 30.XI.16 | 220 158 3½ / 220 198 3½ | 916 457 1373 | 500 800 1000: 6,5 8 / 1500 2000 3000: 13,5 20 40,5 | 80 136 | 16 6 | 40,54 | 33,9 8,43 | 3,84 1,30 0,60 | 12,50 2,80 | 68,7 1400 0,049 | 7,6 5,04 |

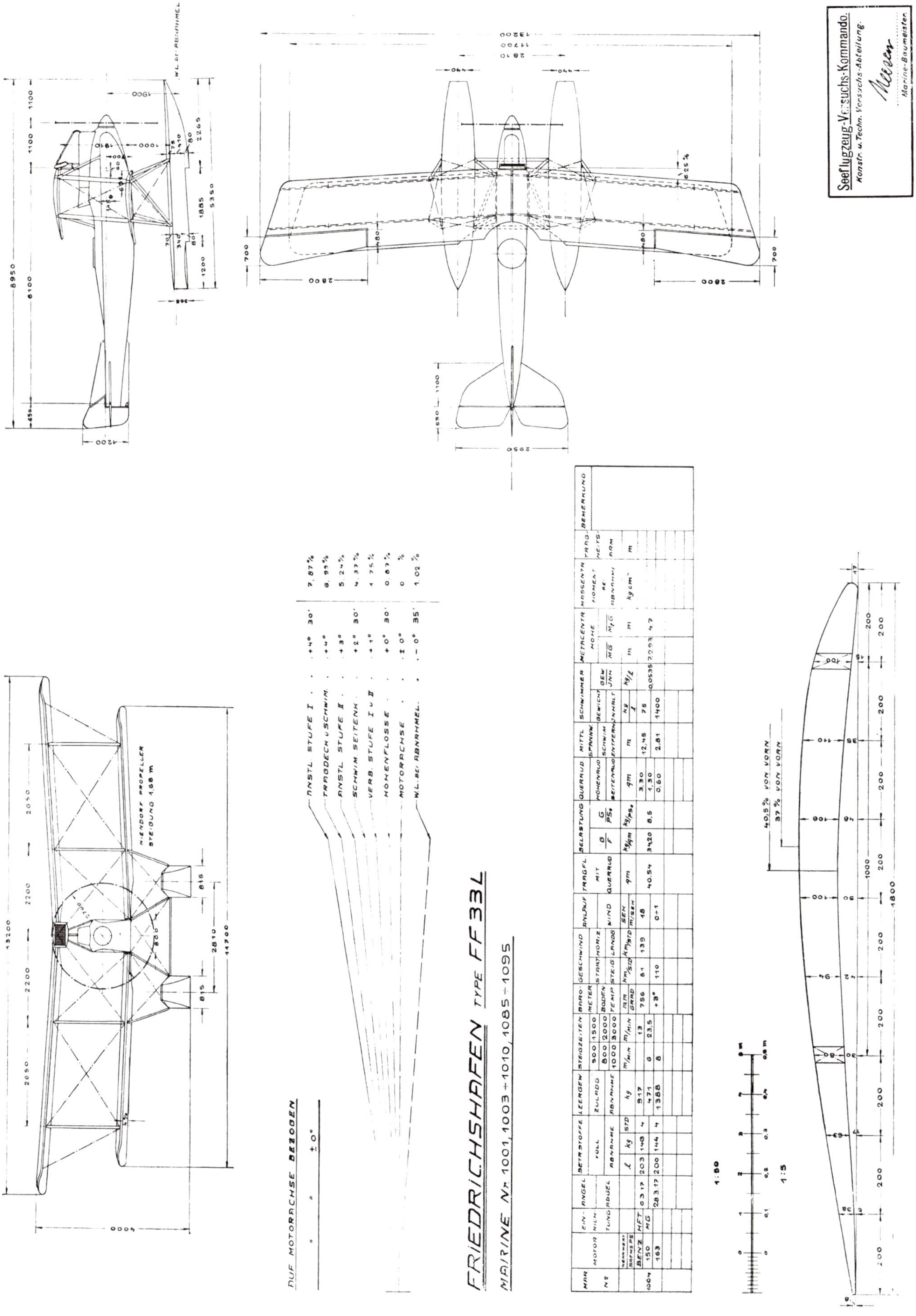
Seeflugzeug-Versuchs-Kommando.
Konstr. u. Techn. Versuchs-Abteilung.
Marine-Baumeister.
FRIEDRICHSHAFEN TYPE FF 33L
MARINE Nr 1001, 1003÷1010, 1085÷1095
NIENDORF PROPELLER STEIGUNG 1,68 m
AUF MOTORACHSE BEZOGEN
405% VON VORN
37% VON VORN
1:50
1:5

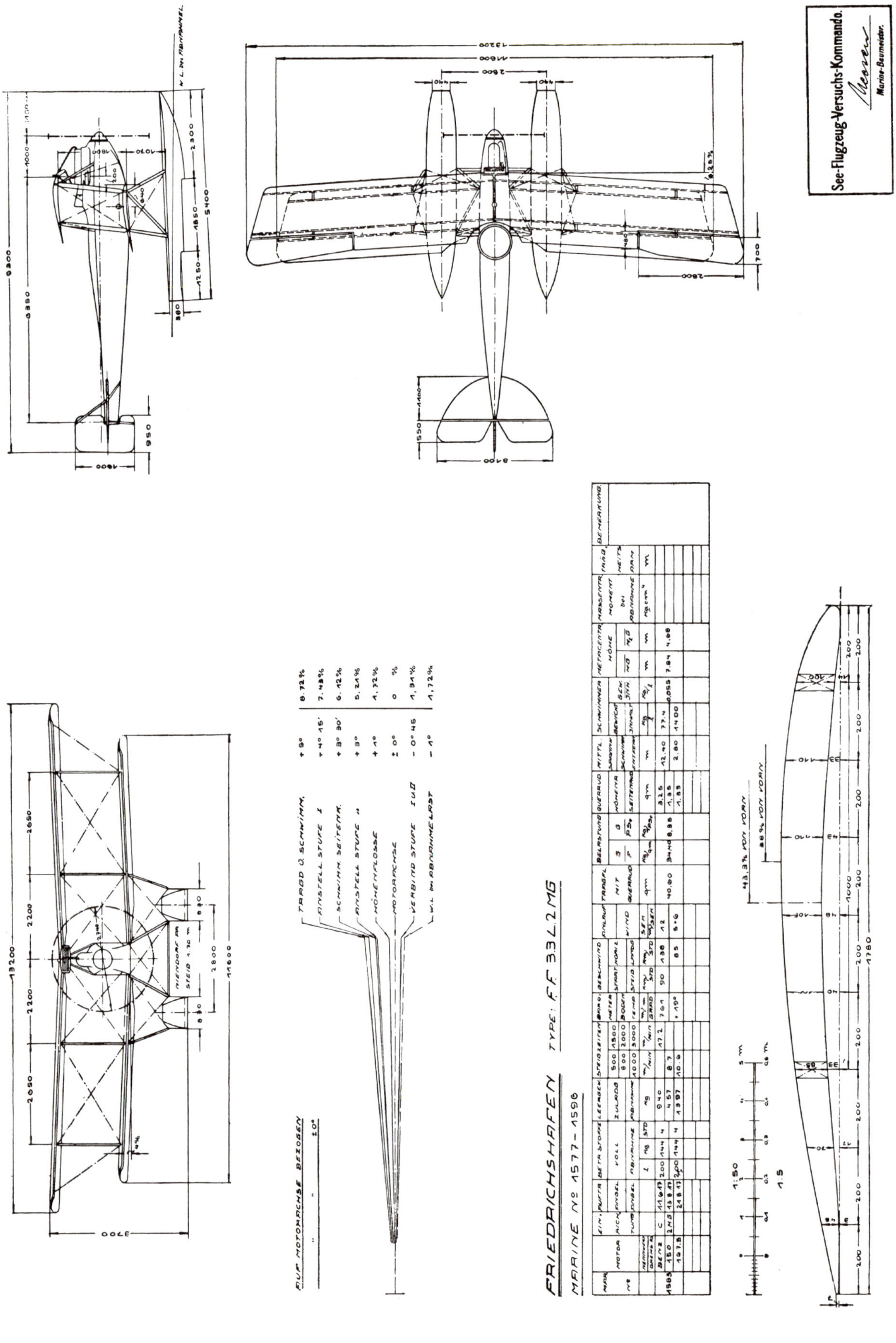
See-Flugzeug-Versuchs-Kommando.
Marine-Baumeister.
FRIEDRICHSHAFEN TYPE: F.F. 33 L 2 MG
MARINE No 1577 - 1596

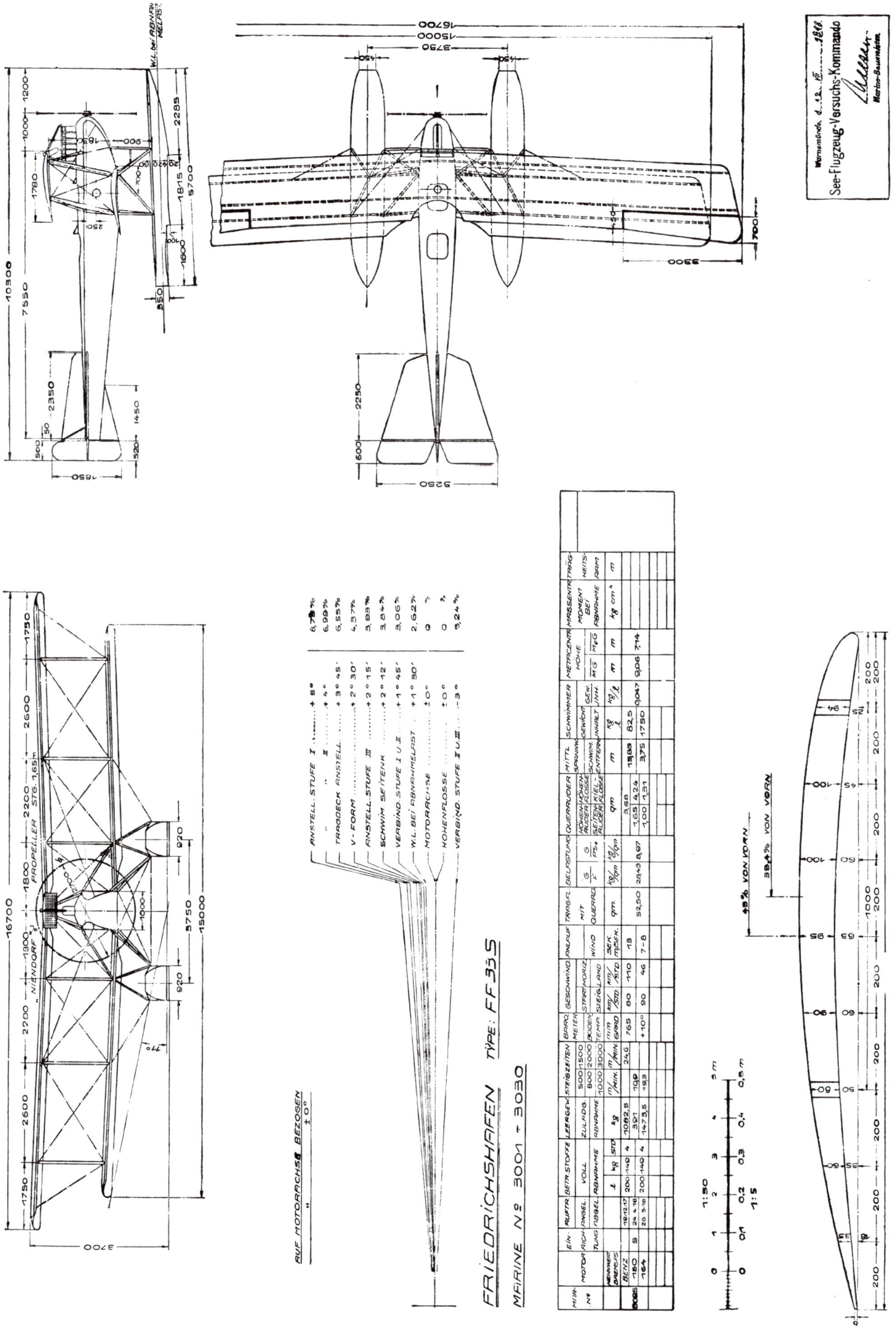
FRIEDRICHSHAFEN TYPE: FF 33 S
MARINE Nº 3001 ÷ 3030
AUF MOTORACHSE BEZOGEN
ANSTELL. STUFE I + 6°
TRAGDECK ANSTELL. + 3° 45'
V-FORM + 2° 30'
ANSTELL. STUFE III + 2° 15'
SCHWIM. SEITENK + 2° 12'
VERBIND. STUFE I u. II + 1° 45'
W.L. BEI ABNAHMELAST + 1° 30'
MOTORACHSE ± 0°
HÖHENFLOSSE ± 0°
VERBIND. STUFE II u. III - 3°
PROPELLER STG. 1,65 m
48% VON VORN
38,4% VON VORN
1:50
1:5
See-Flugzeug-Versuchs-Kommando
Marine-Baumeister

Top: Friedrichshafen FF33E in Bulgarian navy service.

Above: Friedrichshafen FF33H in Bulgarian navy service.

Top: Friedrichshafen FF33E MN 885 in Bulgarian navy service at Varna.

Above: Friedrichshafen FF33H in Bulgarian navy service.

Right: Friedrichshafen FF33 floatplanes in Bulgarian navy service at *Seeflugstation* Varna.

Above: Friedrichshafen FF33L 1271 in Bulgarian navy service at Varna in 1919.

Above: Bulgarian navy insignia.

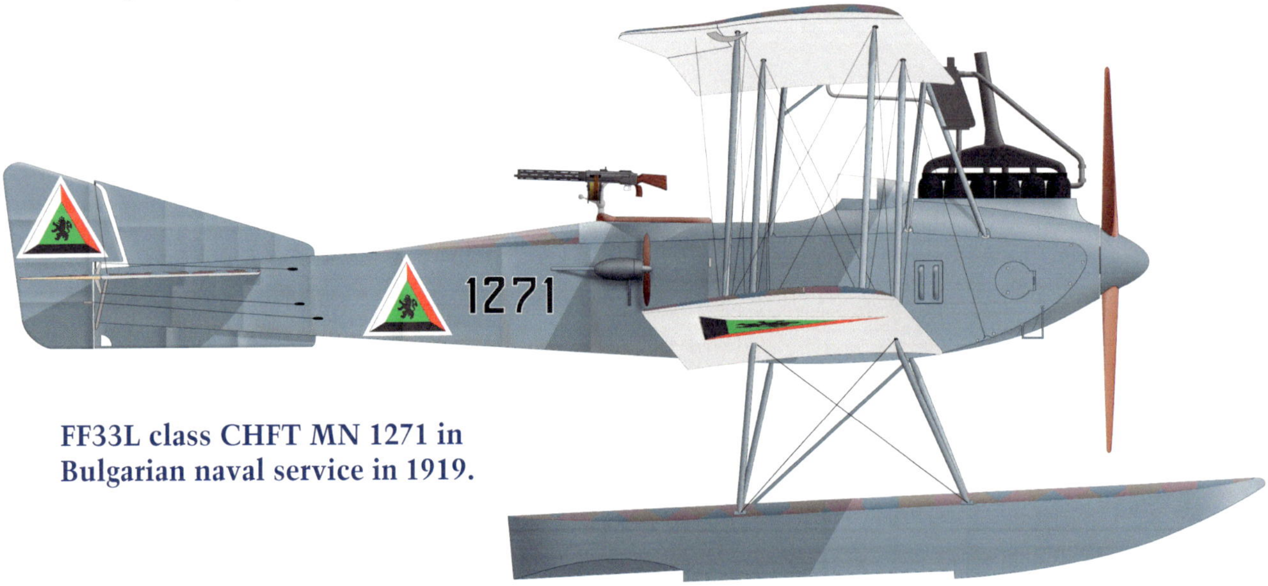

FF33L class CHFT MN 1271 in Bulgarian naval service in 1919.

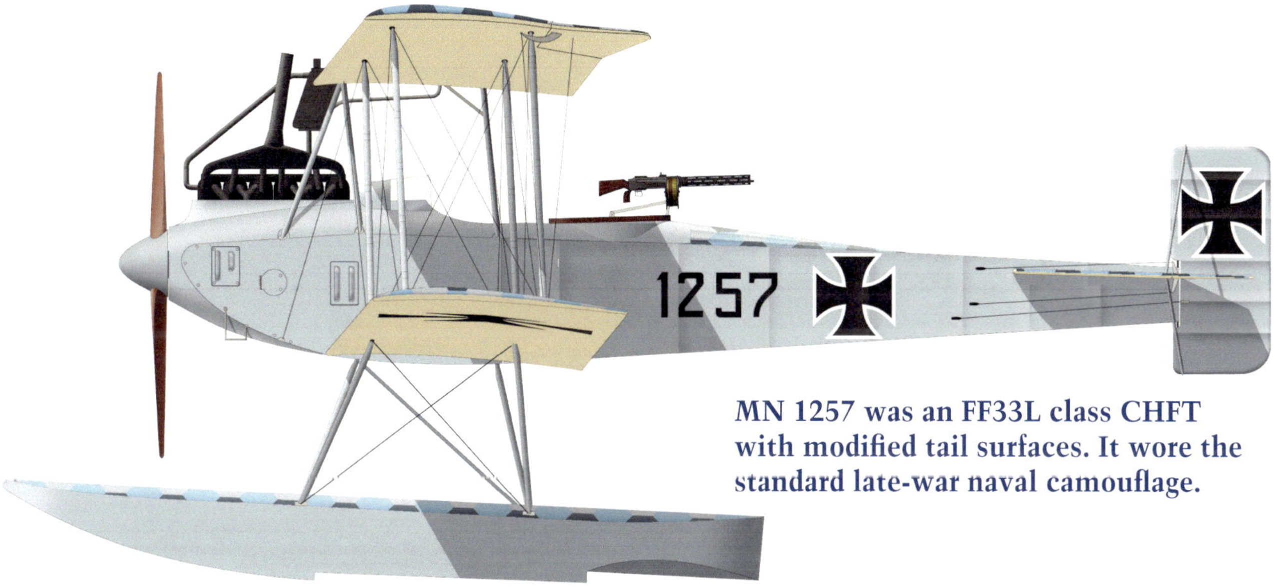

MN 1257 was an FF33L class CHFT with modified tail surfaces. It wore the standard late-war naval camouflage.

MN 3144 was an FF33L class C2MG with experimental tail surfaces. It was finished in late-war naval camouflage with the engine cowling painted darker than standard.

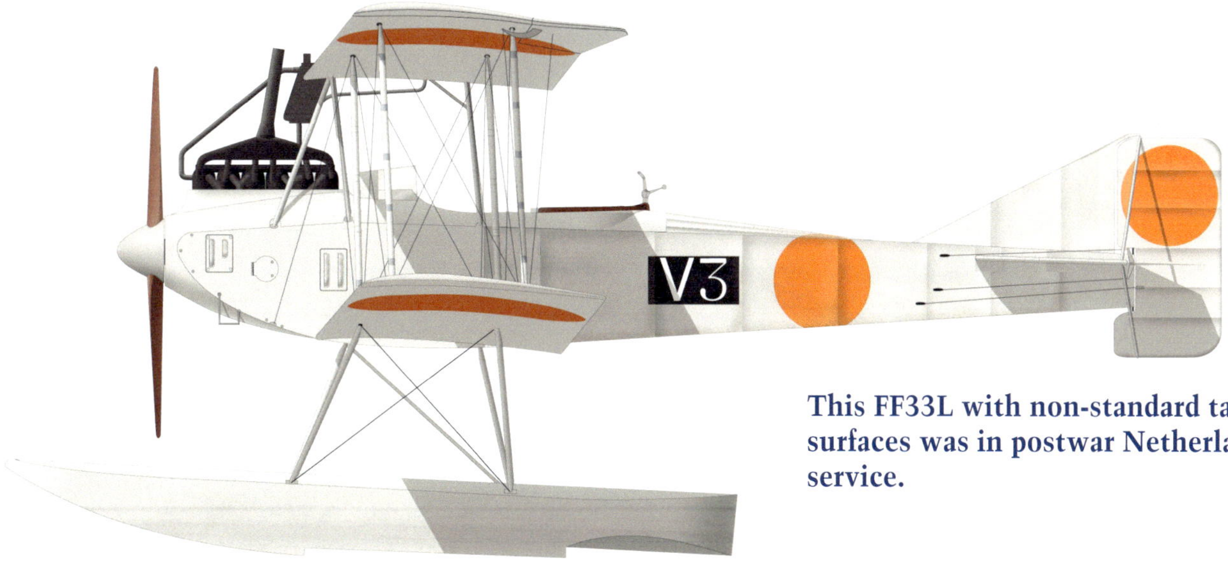

This FF33L with non-standard tail surfaces was in postwar Netherlands service.

Friedrichshafen FF34

Above & Below: The sole prototype Friedrichshafen FF34 was built to explore the pusher configuration for reconnaissance floatplanes.

The Friedrichshafen FF34 was another pusher design to further evaluate the potential of this configuration for the Navy's reconnaissance floatplanes. Built to class CHFT, it was a large aircraft powered by a 240 hp Maybach. To further improve performance compared to the earlier FF31, the FF34 had streamlined booms attaching the tail instead of the complex, drag-producing struts and bracing wires of the FF31.

Delivered to the *SVK* on 24 January 1916, the FF34 proved unsatisfactory. When the *SVK* evaluated the FF34, it determined that the new, streamlined booms were not strong enough and on 22 April the aircraft was returned to Friedrichshafen for rework. Delivered a second time on 4 May 1917, the FF34 was still determined to be insufficiently robust and

Friedrichshafen FF34 Specifications

Engine:	240 hp Maybach Mb.IV	
Wing:	Span Upper	18.40 m
General:	Length	10.85 m
	Height	4.10 m

Above: The prototype Friedrichshafen FF34 first flew in early January 1916 and was delivered on 24 January 1916. Like most pusher designs the observer sat in the forward cockpit where he had an excellent field of view and field of fire for his flexible machine gun. Designed as class CHFT, the FF34 also carried a wireless transmitter and receiver.

it was rebuilt with conventional fuselage and tractor engine to become the prototype FF44. The FF34 was Marine Number 117, and the rebuilt FF44 retained the same Marine Number.

Although the pusher configuration offered an unobstructed field of view and field of fire forward, pushers were too vulnerable to attack from the rear. Between its obsolete configuration and structural limitations, the FF34 remained a single prototype.

Above: This view of the Friedrichshafen FF34 shows the careful attention given to minimize the frontal area to keep drag low despite the pusher configuration.

Above: The prototype Friedrichshafen FF34 at the airship hangar on Lake Constance.

Below: The prototype Friedrichshafen FF34 inside the airship hangar on Lake Constance.

Above: The Friedrichshafen FF34 afloat on Lake Constance. Like all Friedrichshafen aircraft, the FF34 was carefully designed and built using the conventional wire-braced, fabric-covered wood structure of the time. Friedrichshafen's structures were robust but not innovative. Unfortunately, the tail booms were not robust enough for operational use.

Flugzeugbau Friedrichshafen
G. m. b. H. -:- Friedrichshafen am Bodensee

Wasserflugzeuge	**Landflugzeuge**
Lieferant der Marine	Lieferant der Armee

Inhaber von Weltrekorden.

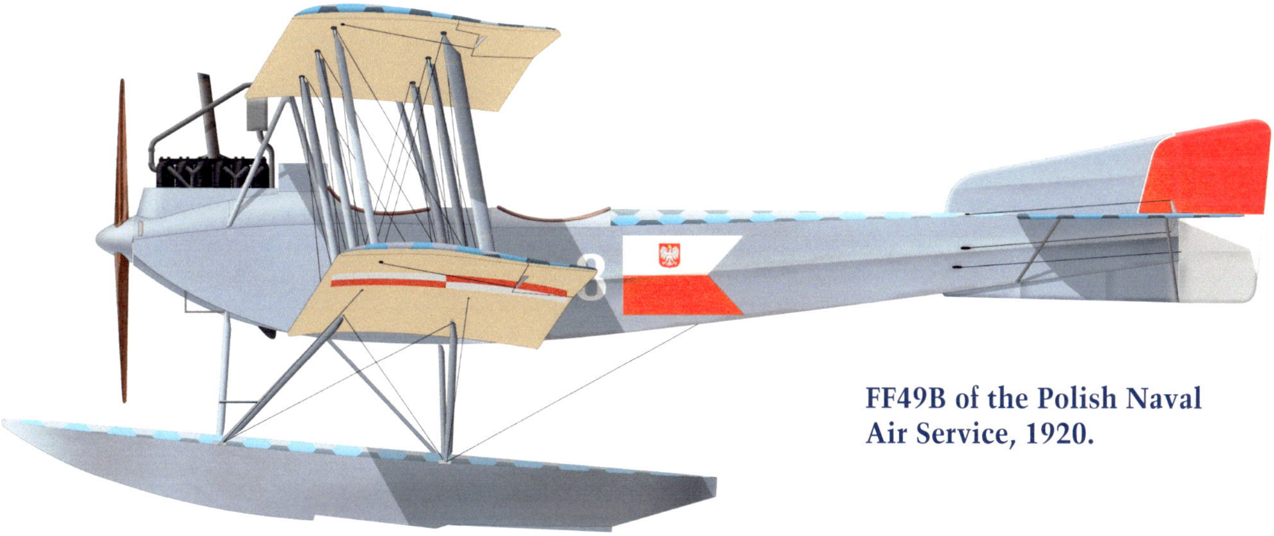

FF49B of the Polish Naval Air Service, 1920.

Friedrichshafen FF39

Above: Marine Number 587 ordered in December 1915 was the prototype FF39. Powered by a 200 hp Benz Bz.IV, the FF39 was a class CHFT floatplane. A total of 14 of these aircraft were built. (The Peter M. Bowers Collection/The Museum of Flight)

Powered by a 200 hp Benz Bz.IV, the FF39 was a larger, heavier, more powerful development of the Friedrichshafen FF33J that used the 150 hp Benz Bz.III. The additional power enabled the FF39 to be armed with a flexible machine gun, whereas the FF33J was not armed.

A full year elapsed between the order for the FF39 prototype and its first flight, most likely due to delayed availability of the required 200 hp Benz engine. The FF39 gave good service but only 14 were built because it was soon replaced in production by the improved FF49C powered by the same engine. The first FF39 was delivered in late March 1917 and the first FF49 aircraft were ordered in May 1917.

Friedrichshafen FF39 Specifications		
Engine:	200 hp Benz Bz.IV	
Wing:	Span Upper	17.10 m
	Span Lower	16.60 m
	Area	68.5 m²
General:	Length	11.40 m
	Height	4.20 m
	Empty Weight	1,438 kg
	Loaded Weight	2,102 kg
Maximum Speed:		137 km/h
Range:		600 km
Climb:	800 m	7.4–9 min
	1,000 m	7–12 min
	1,500 m	11.5–18 min
	2,000 m	17–27 min

Above: The variations in the climbing times given are due to variations in the flying weight due to different equipment and loads carried in the tests.

Below: Prototype FF39 #587.

Friedrichshafen FF39 Production		
Marine Numbers	Quantity	Date Ordered
587	1	December 1915
643–645	3	February 1916
1127–1136	10	December 1916

Above: Ordered in December 1915, first flight of the prototype FF39 was in December 1916 and it was delivered about 25 March 1917. The prolonged development time was probably due to delayed availability of the Benz Bz.IV engine. (The Peter M. Bowers Collection/The Museum of Flight)

Below: The prototype FF39 did not have any wing stagger and featured typical Friedrichshafen lines. (The Peter M. Bowers Collection/The Museum of Flight)

Above: The prototype FF39 displays its clean lines for a reconnaissance floatplane. (The Peter M. Bowers Collection/The Museum of Flight)

Below: Marine Number 643 was the first production FF39. Unlike the prototype, production FF39 aircraft had wing stagger. On 30 September 1917 this aircraft crashed and burned on landing, killing both crewmen. It was attached to Zeebrügge naval air station at the time.

Above: In 1918 FF39 #645, with *Ltn.z.S.* Wolfram Eisenlohr as observer, rendezvouses with a submarine one of Eisenlohr's brothers was serving aboard. Eisenlohr sank the Russian destroyer *Stroini* with bombs while flying FF41AT MN 1000.

Below: FF39 #644 with no spinner wears the late-war camouflage. (The Peter M. Bowers Collection/The Museum of Flight)

This Page & Facing Page, Top: FF39 Marine Number 1127, the first aircraft of the main production batch, wears an interesting, non-standard camouflage scheme. These photos show it with no spinner. (Top photo The Peter M. Bowers Collection/The Museum of Flight)

Above: FF39 Marine Number 1131 of the main production batch taxies on the water. As is evident here, the aircraft of the main production batch were rigged with pronounced wing stagger.

Friedrichshafen FF39 SVK Drawing

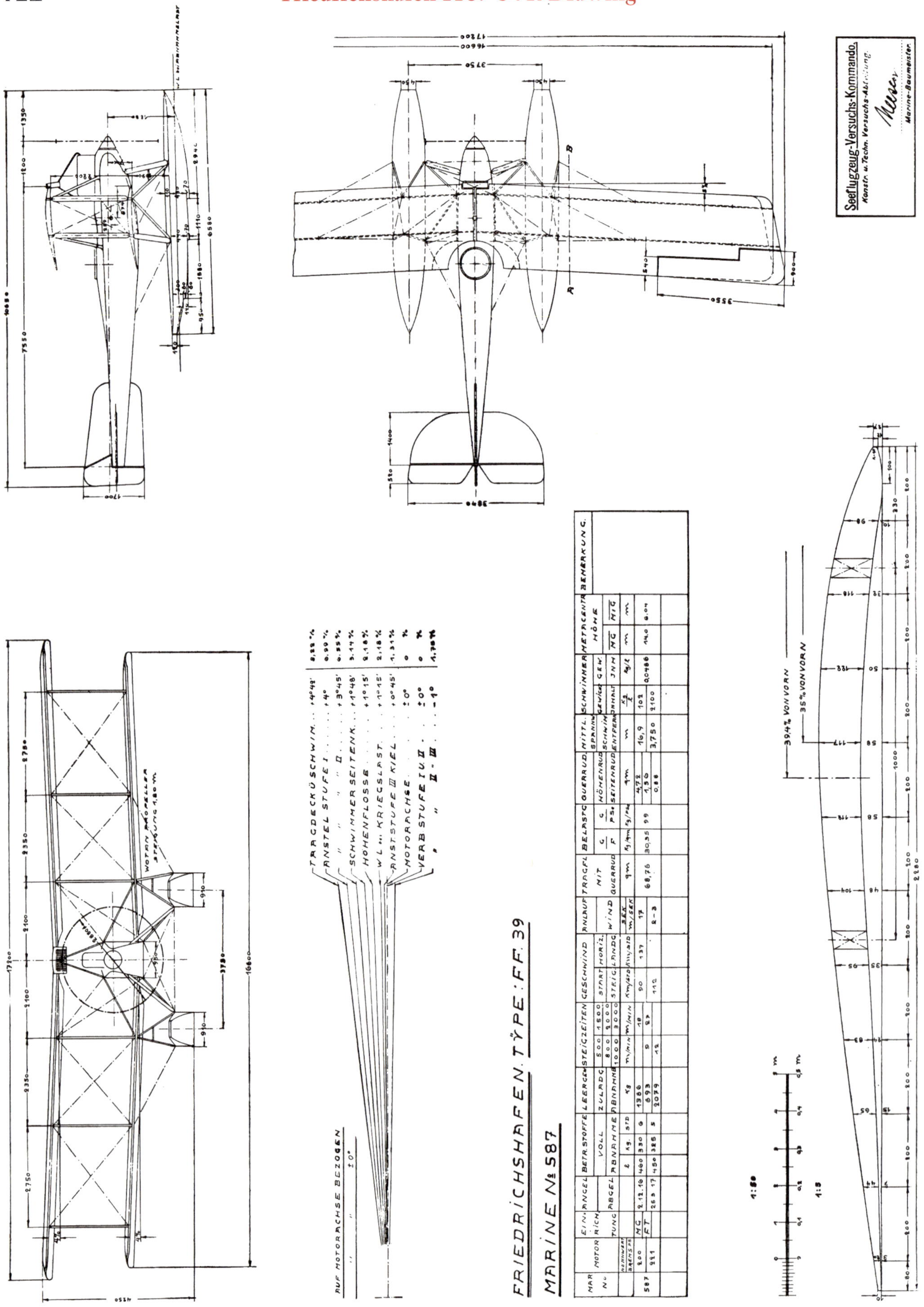

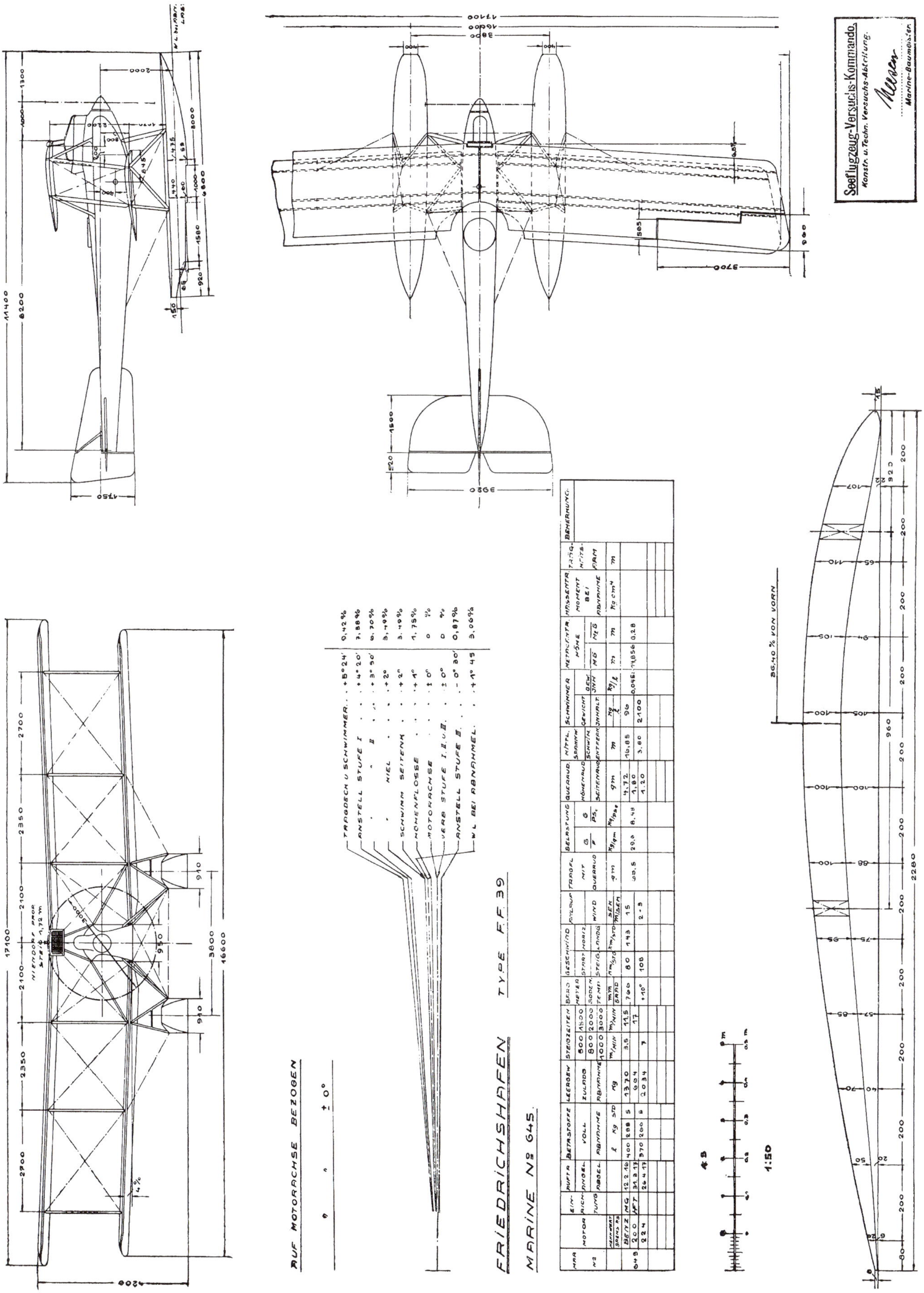
Seeflugzeug-Versuchs-Kommando.
Konstr. u. Techn. Versuchs-Abteilung.
Marine-Baumeister.
FRIEDRICHSHAFEN
TYPE FF 39
MARINE No 645.
1:50

Friedrichshafen FF39 SVK Drawing

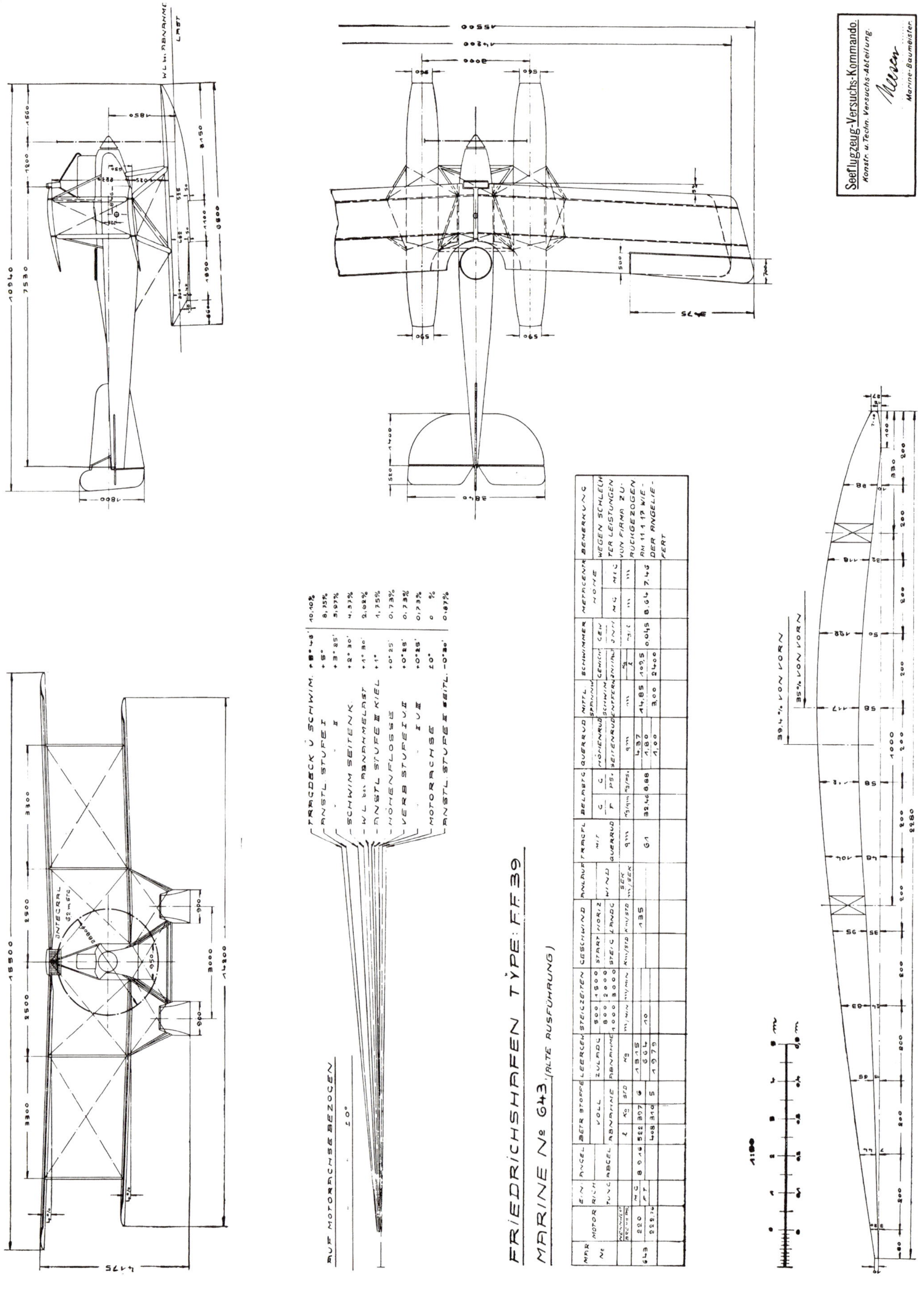

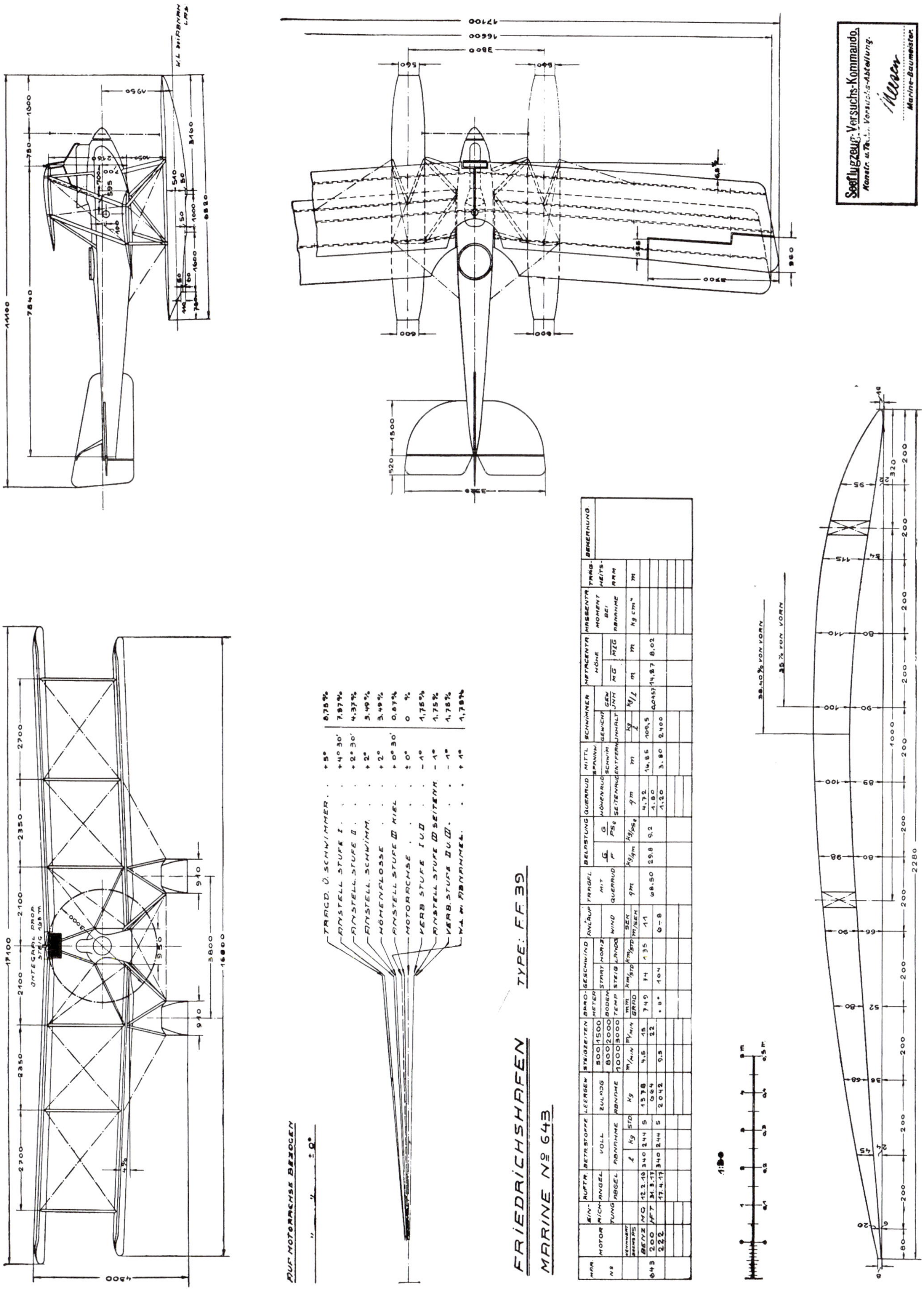
Seeflugzeug-Versuchs-Kommando.
Marine-Baumeister.
FRIEDRICHSHAFEN
MARINE No 643
TYPE: FF 39

Friedrichshafen FF40

Above & Below: FF40 Marine Number 669 shows its unusual configuration of a single engine driving two propellers via ZF bevel gears and propeller shafts, which were enclosed in streamlined fairings. (Top photo The Peter M. Bowers Collection/The Museum of Flight)

Friedrichshafen's FF40 was another prototype to explore to potential of a non-standard configuration, in this case a fuselage-mounted single engine driving two propellers via gears and drive shafts. Like the FF34 and FF44, it was powered by a 240 hp Maybach and was a fairly large floatplane. It was a two-seater, class CFT, with the observer/gunner in the front cockpit, where he had an unobstructed view and field of fire. Unfortunately, the pilot was well aft, making communication and cooperation between the two crewmembers difficult during flight. Marine Number 669 remained the only prototype.

Friedrichshafen FF40 Specifications

Engine:	240 hp Maybach Mb.IV	
Wing:	Span Upper	21.00 m
	Span Lower	19.00 m
	Area	88.9 m^2
General:	Length	12.43 m
	Height	4.45 m
	Empty Weight	1,829 kg
	Loaded Weight	2,539 kg
Maximum Speed:		125 km/h
Range:		650 km
Climb:	800 m	8 min
	1,000 m	10.5 min

Above: The single 240 hp Maybach Mb.IV engine in the FF40 drove two propellers via ZF bevel gears and propeller shafts. The FF40 was ordered in March 1916 and, despite its relatively complex power system, was quickly built and flown. First flight is reported to be in April 1916 with delivery to the SVK in August 1916.

Above: The FF40 was relatively large and powerful for a two-seat reconnaissance floatplane, but the power transmission system was heavy, limiting its equipment to a flexible gun and a wireless transmitter only with no receiver.

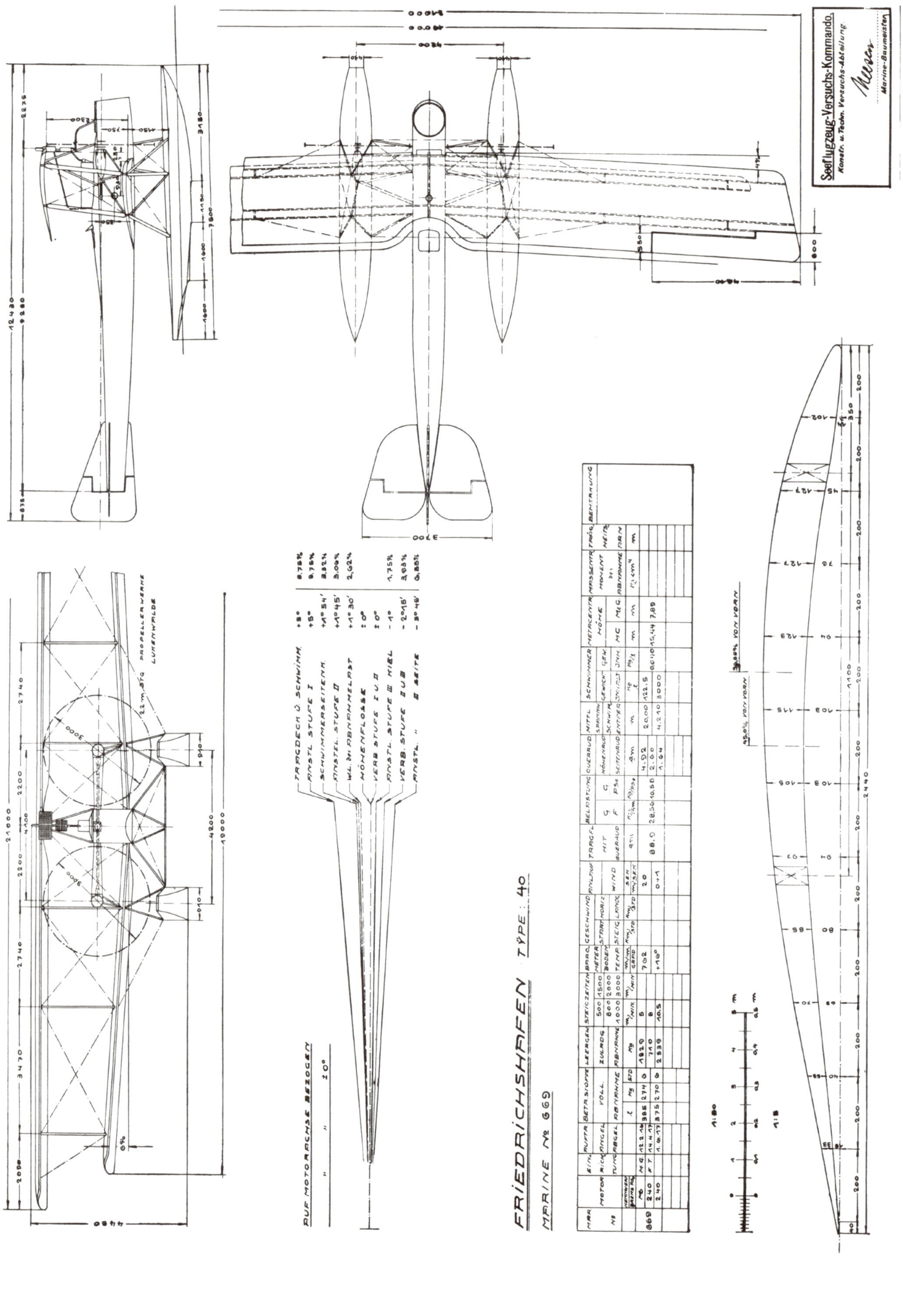
Seeflugzeug-Versuchs-Kommando.
Konstr. u. Techn. Versuchs-Abteilung.
Marine-Baumeister
FRIEDRICHSHAFEN TYPE: 40
MARINE No 669

FF49C MN 1699 in late 1917 featured additions to its late-style camouflage.

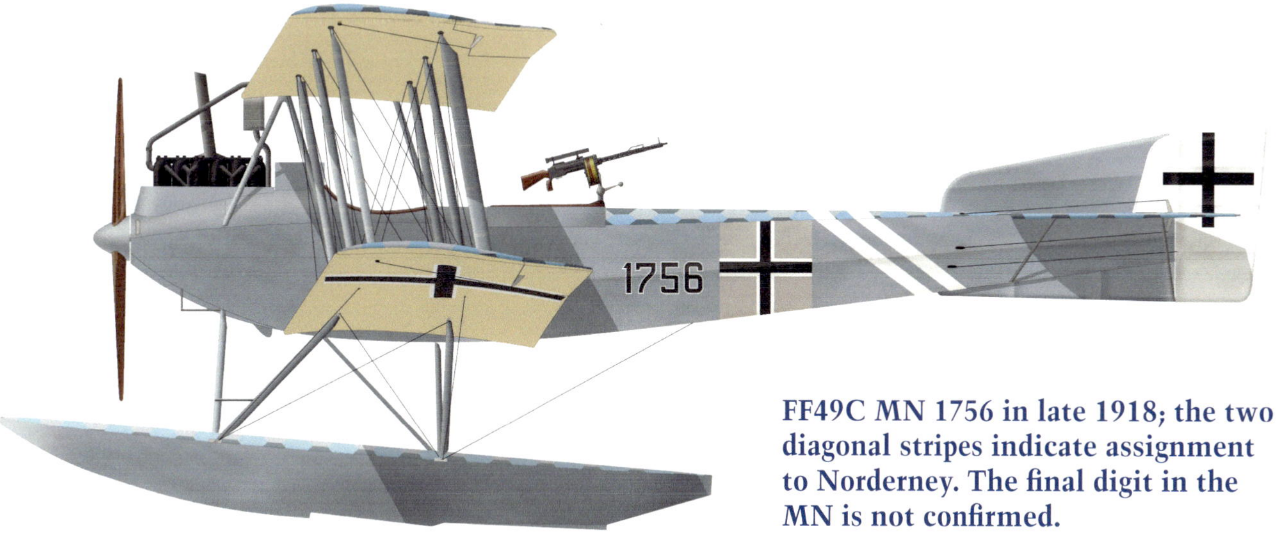

FF49C MN 1756 in late 1918; the two diagonal stripes indicate assignment to Norderney. The final digit in the MN is not confirmed.

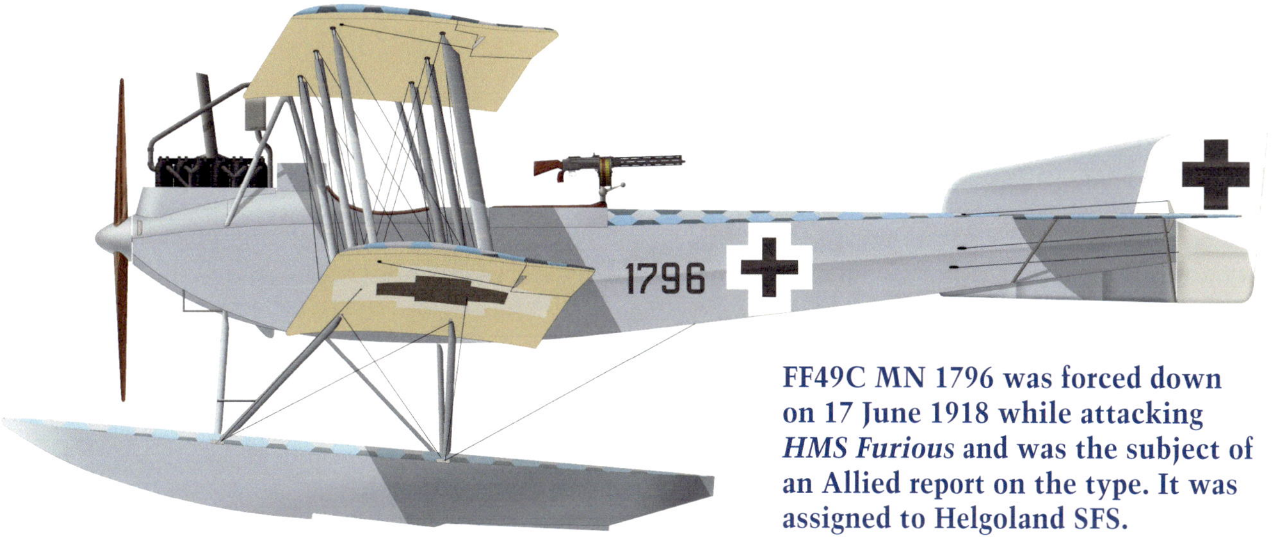

FF49C MN 1796 was forced down on 17 June 1918 while attacking *HMS Furious* and was the subject of an Allied report on the type. It was assigned to Helgoland SFS.

Friedrichshafen FF44

Above: The FF44 was rebuilt from the FF34 pusher and retained its Marine Number of 117.

The FF34 was returned to Friedrichshafen on 22 April 1916 and delivered a second time on 4 May 1917, but was still determined to be insufficiently robust. The Friedrichshafen FF44 was rebuilt from the pusher FF34 using the same wings and engine and retained the Marine Number, 117, that had been assigned to the FF34. Like the FF34, the FF44 was built as a class CHFT.

The FF44 followed typical Friedrichshafen design practice except for its vertical tail surfaces, which featured a balanced rudder without fixed fin, a feature shared with some FF33L aircraft.

To improve propeller efficiency, the FF44 was designed with a reduction gear to reduce propeller speed. The large, slow-turning propeller was expected to improve performance. The FF44 was delivered on 24 October 1917, and this prolonged interval after return of the FF34 was likely due to difficulty getting the propeller reduction gears for the powerful Maybach engine. However, the FF44 did not demonstrate the performance improvement expected and it was cancelled on 31 March 1918, after a single prototype was built.

Friedrichshafen built three two-seat reconnaissance floatplane designs powered by a single 240 hp Maybach Mb.IV, the FF34 pusher, the FF40 with engine in the fuselage driving two propellers via bevel gears and shafts, and the FF44 of conventional configuration with reduction gear. These prototypes were likely built to evaluate and compare these different configurations.

All production Friedrichshafen reconnaissance floatplanes had the conventional configuration of the FF44, clearly indicating its superiority, but no production Friedrichshafen used the 240 hp Maybach. Instead, the much more readily available and significantly more reliable Benz engines, first of 150 hp and later of 200 hp, were used.

Friedrichshafen FF44 Specifications

Engine:	240 hp Maybach Mb.IV	
Wing:	Span Upper	18.40 m
	Span Lower	16.80 m
	Area	68.5 m^2
General:	Length	10.85 m
	Height	4.25 m
	Empty Weight	1,552 kg
	Loaded Weight	2,305 kg
Maximum Speed:		145 km/h
Range:		600 km
Climb:	500 m	6.8 min
	1,000 m	8.9 min
	2,000 m	22.1 min

Above & Below: Rebuilt from the FF34 pusher, the FF44 featured a conventional configuration and retained the Marine Number of 117 first used by the FF34. The FF44 followed the basic Friedrichshafen floatplane formula except for its vertical tail surfaces, which have no fixed fins, just a balanced rudder. This design could also be found on some FF33L aircraft.

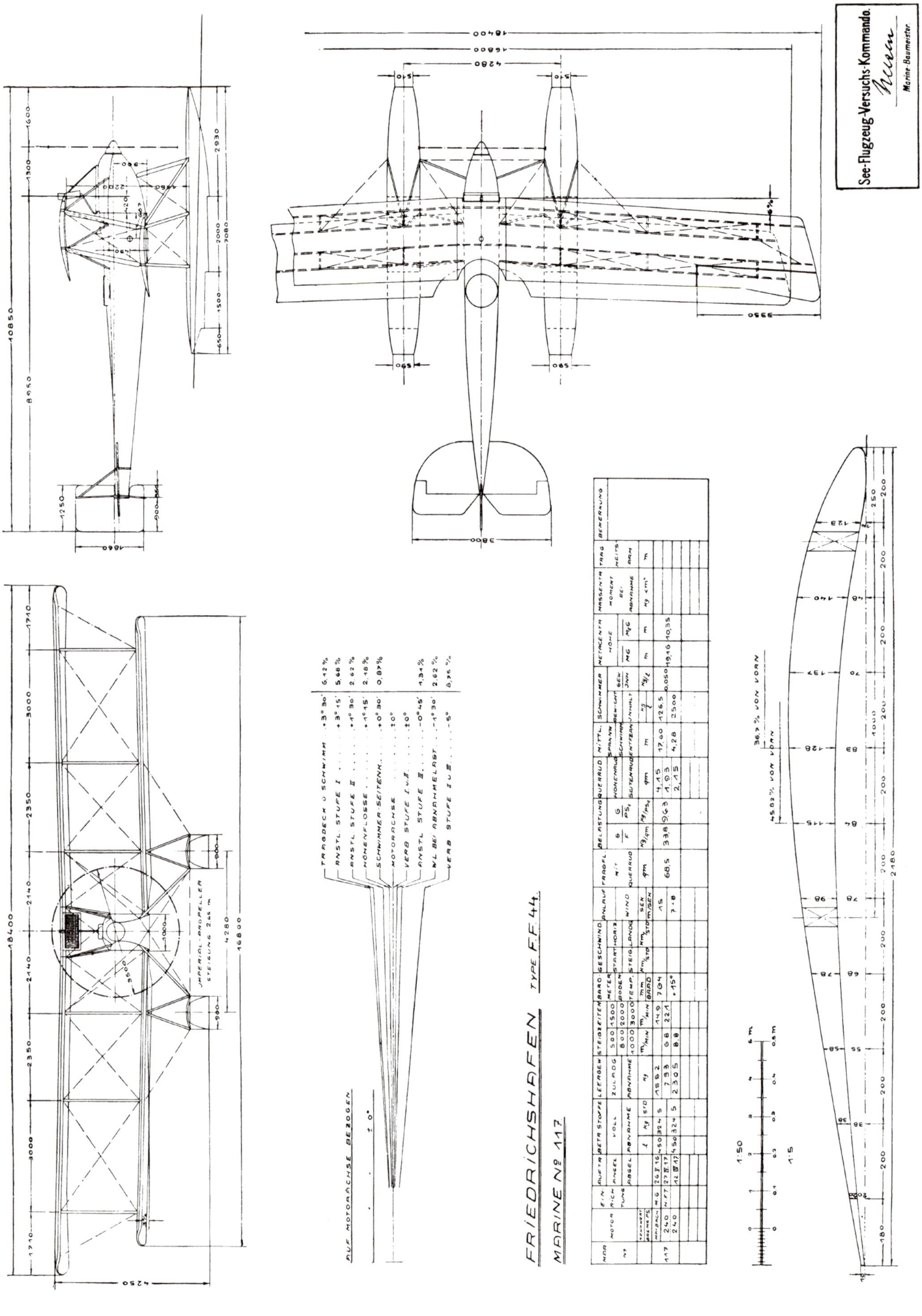
See-Flugzeug-Versuchs-Kommando.
Marine-Baumeister.
FRIEDRICHSHAFEN
TYPE F.F.44.
MARINE No 117

Above: The FF44 Marine Number 117 in the airship hangar at the Friedrichshafen factory. The reduction gear raised the propeller above the engine crankshaft and reduced its speed, allowing for a large, slow-turning propeller that would be more efficient and, it was hoped, provide better performance. The large size of the propeller is evident in this view.

Floatplane Color Profiles

Our talented illustrator has found the Friedrichshafen floatplane profiles the most challenging yet due to the limited amount of information available.

From Alex Imrie's *German Naval Air Service* (Arms & Armour Press, 1989): "Initially seaplanes were not camouflage painted, their natural fabric-covered surfaces being clear doped and varnished; metal parts were finished in grey paint to resist salt water corrosion. However, from April 1917 all top surfaces were to be painted in three colours in hexagonal figures in grey-blue, grey-brown and grey-violet. All side surfaces not visible from above were to be painted grey-blue and the undersurfaces of fuselages and floats painted light grey; the undersurface wing covering remained natural fabric clear-doped finish. Later a printed fabric with a regular hexagon pattern (the top surface colours already mentioned were dyed into the fabric) appeared, and was in widespread use on all top surfaces by the end of the war."

From Heinz J. Nowarra's *Eisernes Kreuz und Balken Kreuz* (Verlag Dieter Hoffman, 1968) discussing early aircraft, pages 53–54: "...the wings of all newly-built aircraft were covered with light grey canvas (linen) which varied in colour since some firms used pre-dyed canvas (linen) and others painted-over canvas (linen). This applied to the fuselages as far as fabric-covered wooden or metal constructions were concerned."

The variations seen in photos of early floatplanes may show experiments in finishes to achieve protection from saltwater corrosion and the sun's ultraviolet rays. The fabric color of the floatplanes could be light gray, light blue, or clear-doped linen, all appearing off-white in photos. In April 1917 naval aircraft camouflage was standardized as noted. During 1916–early 1917 the overall color of floatplanes *may* have evolved into the gray-blue color used as the side color in the late-war standard camouflage.

As a result, the color profiles of aircraft built before the late-war camouflage standard generally must be regarded as provisional.

Friedrichshafen FF49

Above: FF49C Marine Number 1521 was the first Friedrichshafen built. The FF49C was a bigger, more powerful development of the FF33 series to perform the same tasks with better performance and heavier payload. Balanced upper ailerons reduced the pilot's workload on patrols while ailerons on all wings provided better maneuverability.

Special thanks to Colin Owers for contributing this section about the FF49.

The Friedrichshafen FF49 was designed as a successor to the FF33J. The FF39 that followed the FF33J was virtually a refined FF33E and only 14 were supplied from December 1915. The FF49C was a larger aircraft with balanced controls to reduce fatigue on long patrols. It was strengthened to withstand the rigors of floatplane operations and was equipped with the 200 hp Benz Bz.IV motor. The clean lines were complemented by a spinner on the airscrew. The usual ungainly strut arrangement connected the fuselage to the two large pontoons. The observer was equipped with a Parabellum machine gun as standard and both receiving and transmitting wireless equipment were installed. The pilot had a synchronised Maxim 08/15. Of conventional wood and wire construction with fabric covering, the FF49 had two and a half bays to its equal-chord wings. The machine had pronounced sweepback and dihedral. The resulting machine was a robust reconnaissance patrol seaplane.

With a range of some 430 miles the FF49C was well suited to long-range reconnaissance over the North Sea. There were instances where crews that were forced down were rescued by an FF49C, the machine easily taking off with the extra load. Another job for these aircraft was to meet and escort returning submarines through the channels swept in enemy minefields. The FF49C started to replace the FF33 from May 1917 onwards.

The FF49B was a special bomber version of the design wherein the pilot was moved into the rear seat and the observer was not equipped with a machine gun. Although the span of the FF49B was greater than the FF49C, the same area is given for the wings in the Atlas tabulations.

The British captured *MN* 1796 and reported that the machine appeared to be an improved "Friedrichshafoner" D type FF49. The modifications introduced are apparently principally with the object of improving the control and consist in fitting:

(a) Wing flaps on top and bottom wings – the flaps (ailerons – Ed.) being unbalanced, whereas on the earlier type there (sic) *were fitted on the top wing only and balanced.*

(b) Balanced elevators.

(c) small fin underneath the fuselage – the depth of the fin forward being much less than formerly.

(d) Tail plane stays between the bottom of the lower fin and rudder post and the tail plane, instead of above the tail plane.

Friedrichshafen FF49B Specifications

Engine:	200 hp Benz Bz.IV	
Wing:	Span Upper	17.35 m
	Span Lower	17.00 m
	Area	71.16 m²
General:	Length	11.52 m
	Height	4.25 m
	Empty Weight	1,432 kg
	Loaded Weight	2,097 kg
Maximum Speed:		152 km/h
Range:		700 km
Climb:	1,000 m	9 min

Load Carried

These seaplanes carry a crew of two – pilot in front and observer behind in a ring mounting resembling the Scarff. Two machine guns, one firing through the propeller. Six 10 kilo bombs, two of which are submarine bombs and explode on second contact, and six hours fuel.

Wireless both transmitting and receiving as well as a camera are sometimes carried.[1]

1. Group Bulletin, Operations July 1918. TNA AIR1/288/15/226/138.

Friedrichshafen FF49C Specifications

Engine:	200 hp Benz Bz.IV	
Wing:	Span Upper	17.15 m
	Span Lower	16.70 m
	Area	71.16 m²
General:	Length	11.65 m
	Height	4.45 m
	Empty Weight	1,515 kg
	Loaded Weight	2,147 kg
Maximum Speed:		140 km/h
Range:		700 km
Climb:	800 m	6.2 min
	1,000 m	8.0 min
	1,500 m	13.2 min

Netherlands

The Netherlands obtained a number of Friedrichshafen floatplanes that were interned during the war years starting with FF19 *MN* 28 coming down in Dutch waters and washing up on a beach on the Isle of Schiermonnikoog. Stored on the island by the Coast Guard, it was purchased for *f 2,000.* This was the first of a number of British and German marine aircraft that came into the possession of the Netherlands.

Friedrichshafen FF49B & 49C Production Orders

Type	Marine Numbers	Quantity	Date Ordered	Naval Class & Notes
FF49C	1521–1535	15	May 1917	CHFT
FF49B	1536–1550	15	May 1917	BFT
FF49C	1597–1606	10	June 1917	CHFT
FF49B	1607–1609	3	June 1917	CHFT
FF49B	1610–1616	7	June 1917	BFT
FF49C	1669–1698	30	July 1917	CHFT
FF49C	1699–1718	20	August 1917	CHFT
FF49C	1742–1811	70	August 1917	CHFT
FF49C	1812–1841	30	August 1917	C2MGHFT (1822&1823 modified to FF59)
FF49C	1842–1856	15	August 1917	CHFT, built by L.F.G. (Roland)
FF49C	1872–1901	30	August 1917	CHFT, ordered from Sablatnig
FF49C	2053–2092	40	September 1917	CHFT
FF49C	3064–3083	20	June 1918	CHFT
FF49C	3084–3143	60	Unknown	CHFT
FF49C	4004–4033	30	Unknown	C, ordered from Gotha
FF49C	6031–6050	20	Unknown	CHFT, ordered from Sablatnig
FF49C	6051–6080	30	Unknown	CHFT, ordered from Sablatnig
FF49C	6521–6540	20	June 1918	CHFT, built by L.F.G. (Roland)
FF49C	7031–7040	10	Unknown	CHFT, ordered from Travemünde

Note: A total of 25 FF49B & 450 FF49C were ordered. Dates for orders placed after June 1918 are unknown.

Friedrichshafen FF49 Specifications From Different Sources				
	FF49(1)	**FF49B**(2)	**FF49C**(3)	**FF49**(4)
Span, Upper, m	17.350	17.350	17.150	17.15
Span, Lower, m	17.000	17.000	16.700	—
Chord, m	—	2.300	2.300	—
Wing Area, m²	—	71.16	71.16	71.16
Length, m	11.525	11.252	11.650	11.65
Height, m	4.250	4.250	4.450	4.5
Pontoon Length, m	6.555	6.555	6.750	—
Empty Wt, kg	—	1,432	1,515	1,515
Loaded Wt, kg	—	2,027	2,147	2,147
Max. Speed	—	—	—	139.5 km/h
Endurance	—	—	—	5 2/3 hours
Climb to 800 m	—	7.2 minutes	6.2 minutes	—
Climb to 1,000 m	—	9.1 minutes	8.0 minutes	8 minutes
Climb to 1,500 m	—	14.2 minutes	13.2 minutes	—

Source: 1. Friedrichshafen factory three-view. Appears to be FF49B as there is no gun ring.
2. Atlas *MNs* 1536–1550. 3. Atlas MN 1699.
4. Gray, Peter, and Thetford, Owen, *German Aircraft of the First World War*

Three FF49C floatplanes, *MNs* 1679, 1682, and 1705, came into Dutch territorial waters on 3 May 1918. FF49C *MN* 1682 (230 hp Benz No. 31939) from Flanders 1 naval air station was heavily damaged. The observer, *Flugobermatrose* Adolf Haag, later escaped. This machine was purchased for *f* 16,000 and was entered into the *MLD* as V-4.

FF49C *M/N* 1705[2] (230 hp Benz No. 32133) was interned near Wielingen after suffering a motor failure. The machine was severely damaged but the motor had suffered only slight damage. The floats had been damaged by the blows of an axe as the crew, *Flugmaat* Wilhelm Carl Geyger and *Flugmaat* Gerhard Janym from *Flandern* 1, had tried to sink the machine. The aircraft was purchased for *f* 14,000 in February 1919. The floatplane was repaired and displayed at the ELTA exhibition in 1919 with the *MLD* serial V-5.

The seaplane tender *Santa Helena* launched two FF49C floatplanes for a reconnaissance patrol over British waters in the early hours of 10 May 1918. After mapping a new mine field the pair completed their patrol and commenced to return to their vessel. Unfortunately, the aircraft missed the tender and by 11:00 hours and some six hours in the air, they were forced to alight as they were out of fuel.

They sent out an SOS before alighting and set their sea anchors as they expected a rescue vessel to reach them before dark. They were not found by nightfall and a fresh wind sprang up. Shortly after midnight a strong wind broke the sea anchor on *MN* 1874 and it began to drift away. The crew of Hans Sommermann and George Pätzoldt, both NCOs, suffered through a stormy and overcast Sunday and then through another night. They fired a flare every two hours. They were now drifting towards the British shore, but the wind then backed and they started to drift back into the North Sea. The only water they had to drink was that from the radiator and as this was mixed with glycerine, it was hardly palatable. On the fifth day the machine developed a list that the crew unsuccessfully tried to correct by chopping away part of a wing panel but they were too weak. When the fog lifted on the sixth day they saw fishing vessels but they stayed away even though the pair fired flares. Pätzoldt eventually made up a flag with his handkerchief and part of a wing rib and a vessel picked them up. This was the *Argo II*, a Swedish trawler, whose skipper said that he would have approached sooner if they had fired a white flare, but he thought that the red flares meant to stay away. By then the crew only had red flares remaining. They were some 27 miles from the Norwegian coast when picked up and were landed at the Swedish port of Marstrand exactly a week after they were forced to alight. It was calculated that the pair and their redoubtable FF49C had drifted almost

2. WN 698.

Above: FF49B Marine Number 1521 was from the first batch of FF49B aircraft. The FF49B was class BFT, a bomber with no machine gun armament, and the observer was moved forward into the front cockpit. Only 25 FF49B were ordered.

Above: This FF49 aboard ship appears to be FF49B Marine Number 1608 or 1609 from the second batch of FF49B aircraft built. The aircraft is finished in the standard late-war camouflage scheme for German naval aircraft.

Above: A naval aviation unit is lined up for inspection with their FF49C aircraft behind them. The aircraft carry a white chevron unit marking. (Peter M. Bowers Collection/Museum of Flight)

twice across the North Sea.

Their companion FF49C was picked up by a Dutch trawler on the 13th after they had been altered to the situation by an SOS spelt in Morse by bursts of machine gun fire. The Netherlands released the crew they rescued the same month as shipwrecked mariners and the machine was entered into *MLD* service as V-6.

M/N 17585 was interned on 8 August 1918, near Schiermonnikoog, Frisian Islands, after suffering a forced alighting on the morning of the 6th. The crew of *Flugmaat* Willy Poiger and *Flugobermaat* Walter Platow had cut the wings off the machine and sailed it into Dutch territorial waters. This floatplane machine was purchased for *f* 5,000.

1678 crewed by *Flugmaat* K. Gier and *Flugobermaat* Dibrodt from 2e *Seeflug Abt.*, Zeebrügge, was interned on 23 August 1918. Dibrodt later escaped from confinement.

MN 1704, another FF49C was obtained at or around the same time. The crew were *Flugmaat* H. Haus (or Haag?) and *Flugobermaat* Krauthoff, the latter was drowned.

FF49C recorded as 696 crashed on 2 September 1918. This is thought to be its *Werke Nummer* which would mean that it was *MN* 1703. The machine hit a dyke completely wrecking the aircraft. It was identified by plates found in the wreckage. Apparently the motor was purchased.

On 2 October 1918, the two man crew of *MN* 1679, *Flugmaat* Hans Röscher and *Flugmaat* Adolf Imhoff, was picked up by the lugger *Nicolass* and taken ashore at Ijmuiden.

Germany Post-War

Under the Versailles Treaty the German Navy retained aircraft to assist in clearing mine fields. These aircraft were not allowed to be armed in any way. Brandenburg W29 and FF49C floatplanes were the majority of aircraft that operated from Holtenau, Warnemüde, Bug, List, Heliigoland, Norderney, Wilhelmshaven and Hage. In July 1920 the German Government was instructed to remove all mine clearing aircraft from service and deliver them to the Allies.

The German aircraft industry began to convert to civil aircraft almost immediately after the Armistice. A civil register was established that ran from March 1919 to May 1920 when the Allies caused it to be suspended in their attempt to eliminate the German aircraft industry from the post-war world. A second register was established and some aircraft appear in both registers which can be confusing at times. Luft-Fahrzeug-GersellschaftmbH (LFG) was

Above: FF49C #175X carries a double white stripe unit marking. (Peter M. Bowers Collection/Museum of Flight)

Below: FF49C #16X8 is being recovered by hoist. The FF49 was robust enough to withstand rough shipboard handling.

Above: FF49C #189X was one a batch of 30 license-built by Sablatnig, who received three different orders for a total of 80 FF49C floatplanes; all were class CHFT. Damage to the original print obscured the last digit of the Marine Number.

founded in 1908 in Berlin. In order to differentiate its products from rival LVG it adopted the trade name Roland. After the end of the war LFG modified the FF49C floatplanes that it had been building under licence, for civilian use. They had a wider fuselage to allow side by side seating and a new aerofoil section to allow for greater weight to be lifted, and a 220 hp Benz engine. Four aircraft were converted in 1919. The V1 had three passenger seats and an enclosed cabin, the V2 had a longer superstructure with two cabins for five passengers. LFG operated a passenger service from July 1919 with these aircraft from Stralsund to Rügen but this proved difficult to maintain with fuel shortages confining their activities to joy flights on the Baltic coast. The V1 floatplanes named *Max* and *Moritz* received the civil registrations D-132 and D-133[3], while D-242 *Fromme Helene* and D-134 *Witve Bolte* were the V2 version. D-132 and D-242 were impounded by the IAACC in May 1920 but were later released but did not appear on the second German civil register. D-242 was sent on a tour of Sweden in July 1920 but crashed on the return journey.

Deutsche Luft Reederei (DLR), a postwar German airline, used the FF49C as well as a number of other floatplanes in its services. These were given DLR numbers W1, W3, W5 W6, W7 (D-377), W9 D-378), W10 (D-379), W11 (*MN* 3078, D-380), and W12 (D-381).

On 5 July 1920, DLR had the following FF49C aircraft:-

D-378(?) (MN 3076), D-379 (MN 3077) later D-40, D-381 (MN 3079) later D-49, and MN 3080, 3085 to 3091. A total of eleven FF49 C seaplanes.

Sablatnig Flugzeugbau GmbH added passenger cabins to their licence-built FF49C floatplanes and successfully exported a number to Denmark, Norway and Sweden. In June 1921 the Company owned four FF49 floatplanes registered D85, D86, D120 and D121, all fitted with 200 hp Benz engines. Sablatnig operated its airline under the name Luftverkehr Sablatnig from March 1919. This was painted on the fuselage sides in large letters.

FF49C *MN* 1805, FF33E seaplanes *MNs* 1988 and 1990, and FF33S 3012, 3013 and 3015, were preserved for the *Reichmuseum* and along with a number of other aircraft were not required to be destroyed by the IAACC.

Denmark

Denmark's Ministry of Public Works obtained five FF49C floatplanes in 1919 from Germany with the intention of establishing an experimental airmail service. Although the service was successful it was decided to use land-based mail services. The floatplanes were then taken over by the Navy. Given Royal Danish Naval Air Service serial numbers 31 to 35 they served until 1927. The machines were known as the HB II in RDNAS (Hydro Biplane II) and returned to their reconnaissance role.

Sablatnig, like many German manufacturers, tried to send as many aircraft outside the reach of the

3. D133 was a V2 according to D-242.

Above: FF49C #1699 was first of a batch of 20 (1699–1718) ordered in August 1917 and carried a checkerboard marking.

Below: FF49C #1602 was one of a batch of 10 (1597–1606) ordered in June 1917 and carried standard late-war camouflage.

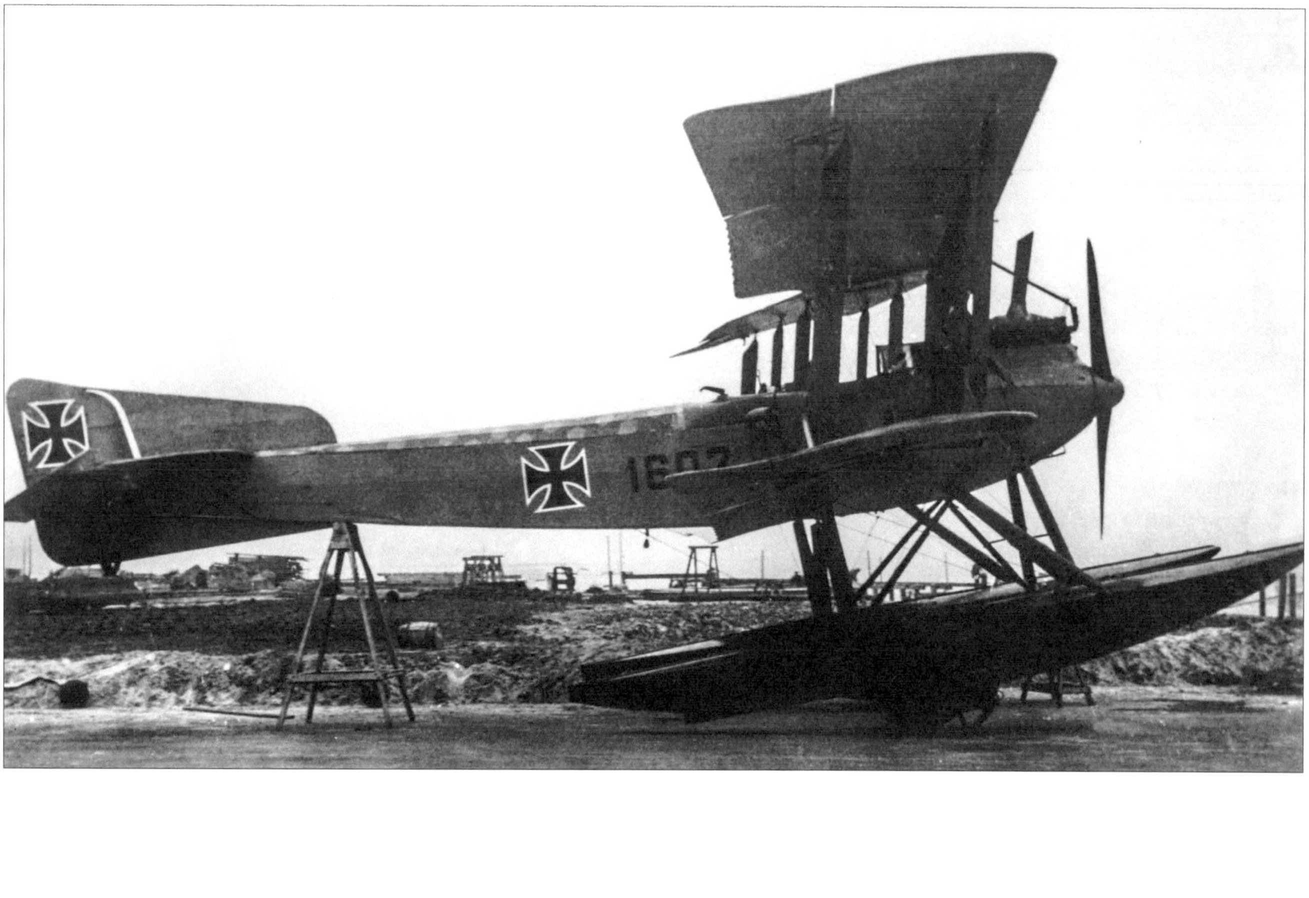

Above: Nose details of FF49C #1699 that carries a checkerboard marking; the entire nose is painted dark (black?).

IAACC as possible and sent four FF49C floatplanes to Denmark for use by Dansk Luft-Express but were taken in hand by Denmark's custom service and stored at Aabenraa in January 1919. These four were obtained by the Danish Navy in April 1922. Only two actually entered service, the other two were used for spares. They received the serials 36 and 37.

The first flights of the Danish airline Det Danske Luftfartselskab were carried out from 7 August by a FF49C obtained (or leased) from the German airline Deutsche Luft-Reederi (DLR). This was registered T-DABA. Apparently this acquisition was not permitted under the Peace treaty and the machine was returned to Germany to be replaced by another FF49C, *MN* 3078, and this seaplane was also registered T-DABA. This second aircraft was apparently constructed from the fuselage of the original aircraft that was taken back to Denmark and spare parts. Registered in June 1921 the venture proved a financial disaster and the machine was obtained by the RDNAS as spares for its own Friedrichshafens in late 1921, the registration being cancelled the following year.

Three Danish FF49C floatplanes arrived at Horten after a non-stop flight from Copenhagen on 19 September 1920. An unusual feature of this arrival was the release of pigeons carrying messages back to Denmark. The use of carrier pigeons was a feature of marine aircraft in the recent war with the birds being responsible for the rescue of crews whose seaplanes had come down on the North Sea.

A static reconstruction of an FF49C is presently being undertaken for the Danish Aviation Museum.

Finland

A civil war in Finland broke out in 1918 between the Bolshivek (Red) and White forces. Germany sent troops into Finland in early April 1918 at the request of the Finnish government. It suited the Germans to stop a new front opening with the Soviet government. They established a naval base at Koivisto with three Friedrichshafen seaplanes. FF49C *Marine Nummer* 1799 operated an air-mail service in September 1918 from Hermanni air station until it was destroyed in a crash in the bay just outside the station. After the Armistice the Germans were forced to withdraw from Finland in December 1918 and they took all their equipment including all aircraft with them.

The end of the war saw Finland looked upon as a bulwark against the Reds and the French sent an aviation mission to Finland early in 1919. In the

Above: FF49C #1832 was one of a batch of 30 (1812–1841) class C2MGHFT ordered in August 1917 and carried the late-style national insignia and standard late-war naval aircraft camouflage.

Below: FF49C wearing the late-style national insignia. Later production FF49C aircraft had an actuating strut connecting their ailerons making them more robust than early production FF49C aircraft that connected their ailerons with a wire.

Above: FF49C #1872 was the first aircraft of a batch of 30 (1872–1901) ordered from Sablatnig in August 1917.

meantime the Finns had obtained aircraft from other sources that included German and Russian types. Friedrichshafen FF41A and FF49 seaplanes were included in the mix.

Finland obtained the following Friedrichshafen FF49 biplanes:

FF49B *MN* 1546 that received the Finnish serial C84/18 (later 2B307). Purchased from the German Naval air station at Tallinn, Estonia on 26 November 1918. Sank at Koivistolla, 15 October 1919.

FF49C *MN* 1749 received the serial F18 (later C59/18). Was purchased from Germany in 1918. Deleted 11 August 1919 after it was written off in a fatal accident that month.

FF49C *MN* 1750 received the serial F22 (later C72/18 and 2A300). Was acquired from Germany in 1918. Deleted 4 April 1923 with a total flight time 114:17 hours.

FF49C *MN* 1807 received the serial C82/18 (later 2B306).). Obtained from the German Naval air station at Tallinn, Estonia on 26 November 1918. Survived at least until January 1919.

FF49C *MN* 1810 received the serial C85/18 (later 2C302).). Obtained from the German Naval air station at Tallinn, Estonia on 26 November 1918. Deleted 25 May 1923 with a total flight time of 216 hours.

Norway

In the spring of 1920, the new Norwegian airline company Aero A/S started a co-operation with A/S Norsk Aeroplanfabrik and it appears that this company was the one that ordered the five aircraft mentioned in the newspapers. Four of these machines were FF49C floatplanes. Three arrived in late February and were converted into a civil version at A/S Norsk Aeroplanfabrik. The fourth crashed on its delivery flight. They received the civil registrations N6, N7 (Changed to N9) and N8. N6 was registered on 1 May 1920, in the name of Aero A/S and was in June 1920 used on the first charter flight in Norway, when it carried newspapers between Kristiania and Kristiansund on behalf of the newspaper "Nationen". All were leased to Det Norske Luftfartsrederi A/S (DNL). After a minor accident N6 was transferred back to Tønsberg for repairs and storage. N6 was eventually sold to the Army in 1922 which used it for spares. N8 crashed and was destroyed beyond repair after an accident near Haugesund. Its registration was subsequently cancelled. The fourth FF49C was flown to Norway by the German pilot Willy Nolting but he crashed destroying the machine and seriously injuring himself at Copenhagen on 15 April 1920.

Above: FF49C #1878 was one of a batch of 30 (1872–1901) ordered from Sablatnig in August 1917 and carries the early-style national insignia and standard late-war naval aircraft camouflage. No actuating struts connect the ailerons.

Below: An unknown FF49C wearing the late-style national insignia and standard late-war naval aircraft camouflage.

Above: FF49C in Netherlands service postwar; the crewman provides scale for this large aircraft.

JL Tidemanns Tobaksfaabrik whose famous Brandenburg W.33 monoplane (N.21) was used for advertising had earlier obtained a FF49C when Aero A/S was wound up. This was in use from November 1920 to the spring of 1922 when it was written off after a crash.[4] Its replacement was the W.33.

Despite the good service of the Friedrichshafen floatplanes in Norwegian civil flying, none were operated by the Norwegian military.

Poland

Poland bought one FF33E, one FF33L and one FF49B from Danzig in August 1920, and they were delivered to the Polish Navy.

Russia

At least one FF49C was flown in Imperial Russian markings. The White Russian forces took over a number of marine aircraft of German origin in early 1919 including between six to twelve FF49 floatplanes types B and C along with Albatros W4 and Sablatnig SF5 floatplanes.

The short lived Crimea Republic operated FF49C floatplanes *MN* 1554, 1683, 1715, 1716 and 1809 along with a number of FF33 series, Albatros W4, Brandenburg W13, GW and Sablatnig SF5 seaplanes.

Sweden

Two FF49C floatplanes (*MNs* 3064 and 3068) were purchased from Friedrichshafen in January 1919. They were powered by 220-hp Benz engines. They received Swedish Navy serials 23 and 24. Both were written off in 1924.

Svensk Lufttrafik AB obtained two of the seaplanes from Sablatnig that had been converted to carry passengers, however the venture failed and they were also taken over by the Navy. Reconverted to military configuration they received serials 25 and 26. No.25 was used for torpedo trials in 1923. They were to be fitted with dual control for use as trainers. The Swedish Friedrichshafens were never armed. Sweden had also obtained five FF33 seaplanes in August 1918 and the Navy also built five copies as the Sk2 from 1928.

Yugoslavia

Obtained from Austro-Hungarian aircraft depot at Kotor in Montenegro was a single FF49C (*MN* 1833) obtained by the Yugoslavian naval air service. Serialled N2-D it carried the code H12 on the fuselage that also bore a white fuselage stripe indicating it was a training machine.

4. Another reference states that the machine was sold to a J.B. Stang of Oslo. The machine's fate remains unknown.

Right: FF49C V2 in postwar Netherlands service. No actuating struts connect the ailerons.

Below: Front view of an unknown FF49C in postwar Netherlands service. The propeller thrust line is slightly offset to the right to compensate for "P factor" (asymmetric propeller thrust). The downward moving blade has higher angle of attack and thus more 'lift' (thrust), causing "P factor".

Bottom: FF49C V4 ready to launch in postwar Netherlands service.

Above: FF49C V6 in Netherlands service postwar is the center of attention for an avid photographer.

Above: In 1919 Roland modified four of the FF49C aircraft they were building into passenger aircraft. *MAX* was a type V1 which had a widened fuselage with cabin for three passenger seats, 220 hp Benz engine, and modified airfoils for greater lift. Here *MAX*, given civil registration D-132, forms the background for a portrait of people at a beach party.

Belgium

IAACC ordered that six FF49 floatplanes be delivered to Belgium in March 1920. These had been used from the naval air station at Heligoland and Holtenau for mine spotting and search operations. Although delivered they were never used as Belgium had no naval air unit.

Right: FF49C #33 in postwar Danish naval service.

Above: FF49C MN 1796, the captured FF49C described in the British report on p.134, awaiting rescue. FF49C MN1796 was captured after being shot down when it attacked British aircraft carrier *HMS Furious* in company with other aircraft on 17 June 1918. MN 1796 was operating from List on Slyt. *Furious* was engaged in a reconnaissance of the Heligoland Bight before the Tondern raid. *Furious* launched a Ships Camel piloted by Lt. G. Heath who brought down FF49C #1796. The crew of *Leutenant der Reserve* Wenke and *Flugzeugmaat* Schirra were rescued before the aircraft was sunk by gunfire. (See *Royal Naval Air Service 1912–1918* by Brad King, p.117.)

Right: Wreckage of an FF49C on top of the Officer's Casino at Kiel-Holtenau shows the late-war camouflage fabric on the upper surfaces. Unfortunately, we do not have the pilot's explanation for this accident to the authorities, which must have been fascinating.

FF49C W7 in postwar Deutsche Luft Reederei (DLR) service. German airline DLR used 11 FF49C aircraft; W7 was later registered D-377.

FF49C in postwar Finnish service. MN 1810 was acquired from the former German base at Tallinn, Estonia on 26 Nov. 1918 and given serial 85/18, later changed to 2C 302 as shown here. It was deleted 25 May 1923 after 216 hours total flight time.

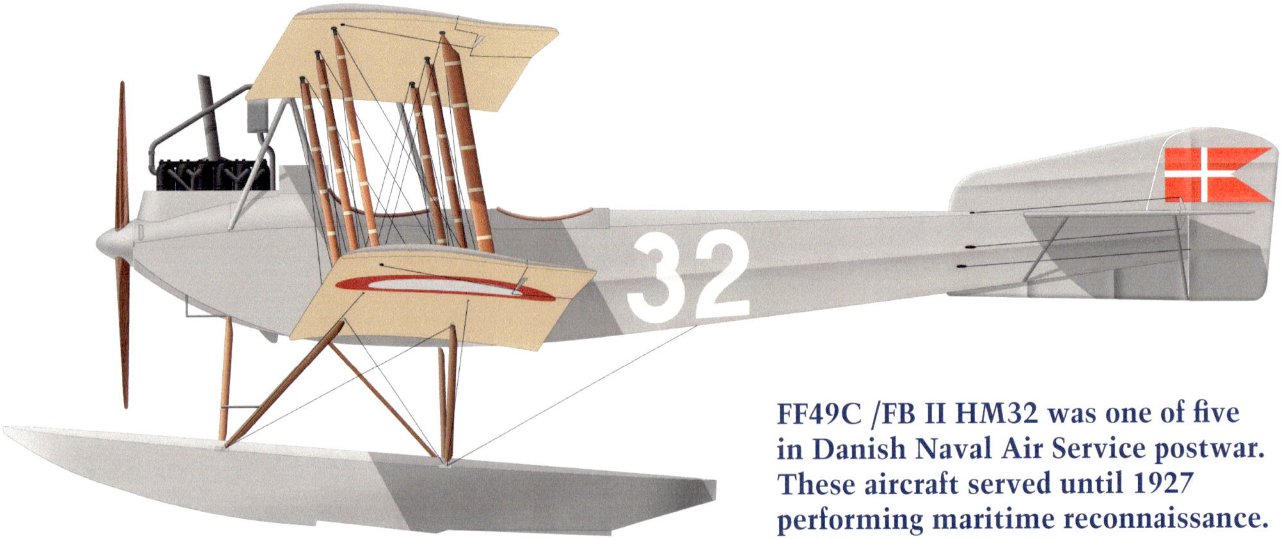

FF49C /FB II HM32 was one of five in Danish Naval Air Service postwar. These aircraft served until 1927 performing maritime reconnaissance.

Above: FF49C type V1 passenger aircraft conversion *MAX*. *MAX* was given civil registration D-132 and seated three passengers in its cabin.

Below: FF49C type V2 passenger aircraft conversion *Witve Bolte*, which was given civil registration D-134. The two V2 conversions had a longer passenger cabin (divided into two) and seated five passengers.

Above: FF49C type V1 passenger aircraft conversion with three passenger seats. This might be *MAX*, civil registration D-132, but the name is not visible; this type V1 might also be *Moritz*, civil registration D-133.

Above: FF49C at Friedrichshafen's Manzell production facility. At left is Eugen Schramm, who was historian Michael Schmeelke's grandfather. Herr Schramm was badly wounded at the Somme in 1916, losing one eye and a finger. After leaving the army for medical reasons, he went to work at Friedrichshafen's Manzell facility. In 1919 he moved to the Maybach Works in Friedrichshafen. Unfortunately, the other workers are unidentified.

Above: FF49C "W7" (D-377) in DLR passenger service in postwar Germany.

Above: This FF49C in postwar passenger service was covered with camouflage printed fabric. (Peter M. Bowers Collection/Museum of Flight)

Above & Above Right: FF49C floatplanes in postwar Danish service.

Above: FF49C T-DABA was in postwar Danish passenger service.

Above & Below: FF49C aircraft in postwar Danish naval service where they were used in their original reconnaissance role.

Above, Top, & Right: FF49C T-DABA in postwar Danish passenger service. This FF49C aircraft was MN 3078. This aircraft has the aileron actuating strut of later production aircraft consistent with its Marine Number.

Above: FF49C #31 in postwar Danish naval service.

Above: FF49C in postwar Finnish service.

Left: FF49C "2C 302" in postwar Finnish service.

Above: FF49C in postwar Russian service.

Left: FF49C in postwar Polish service.

Above: FF49C "26" in postwar Swedish naval service. The extensive postwar service with numerous different air arms was a testament to the good performance, reliability, and robustness of the FF49's design, and also makes its license production more understandable.

Left: FF49C "8" in postwar Polish service.

Above: FF49C in postwar Polish service.

Above: FF49C in postwar Polish service.

This Page & Facing Page: FF49C reproduction being built by the Danish Technik-Museum shows the structure in detail. All Friedrichshafen aircraft featured this type of construction, which was the typical state of the art during WWI. (Photos courtesy F. Gerdessen)

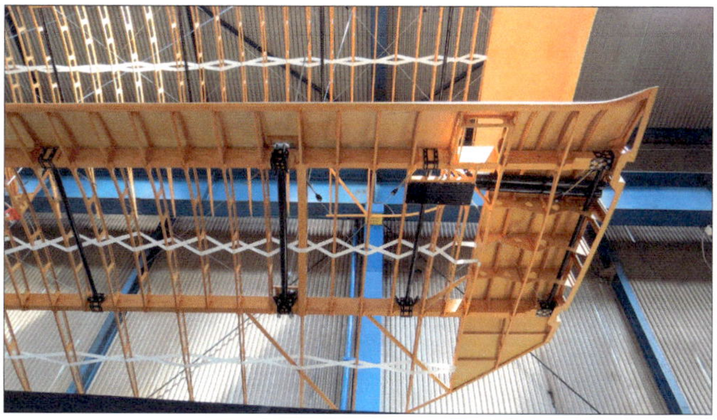

Friedrichshafen FF49B SVK Drawing

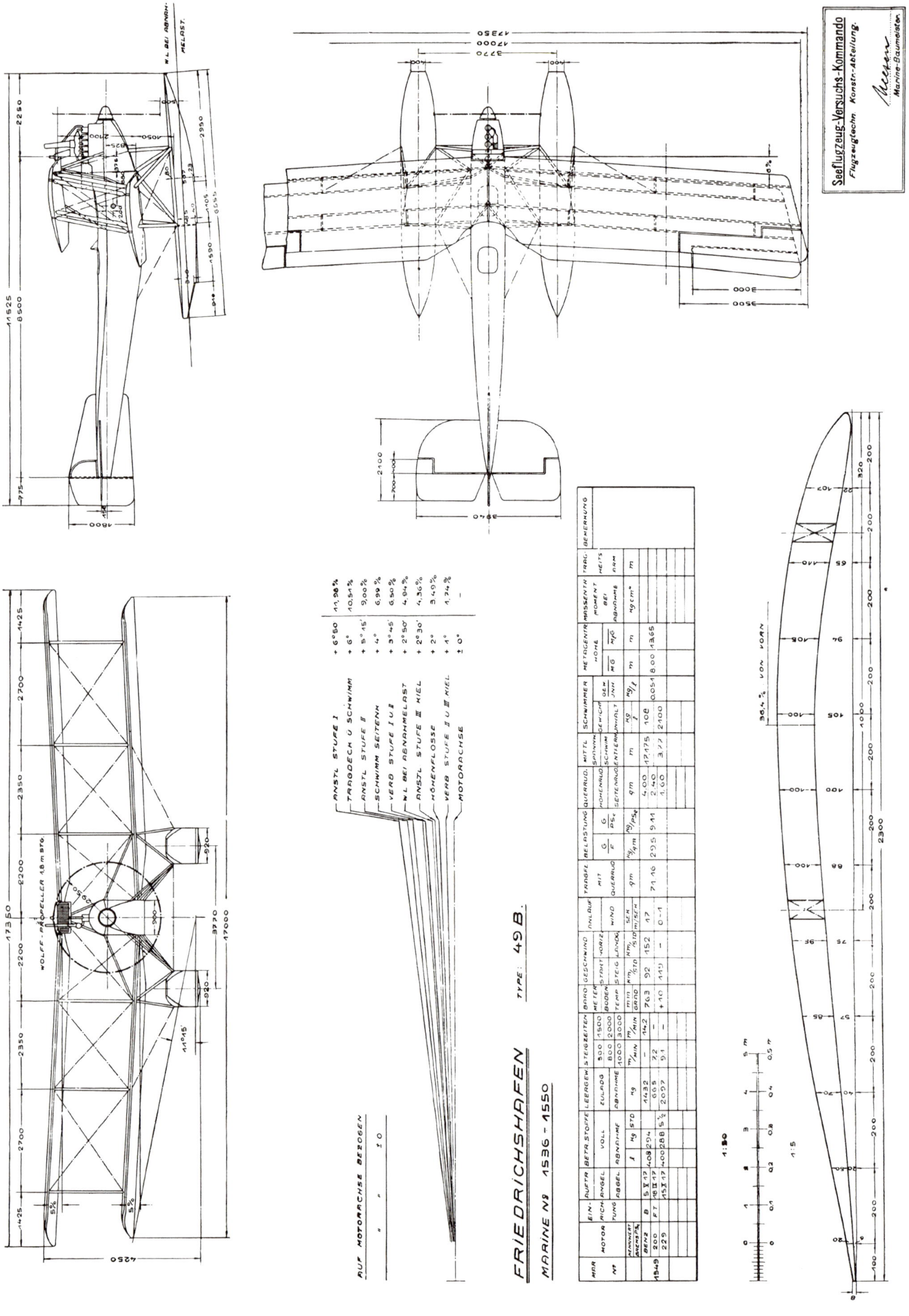

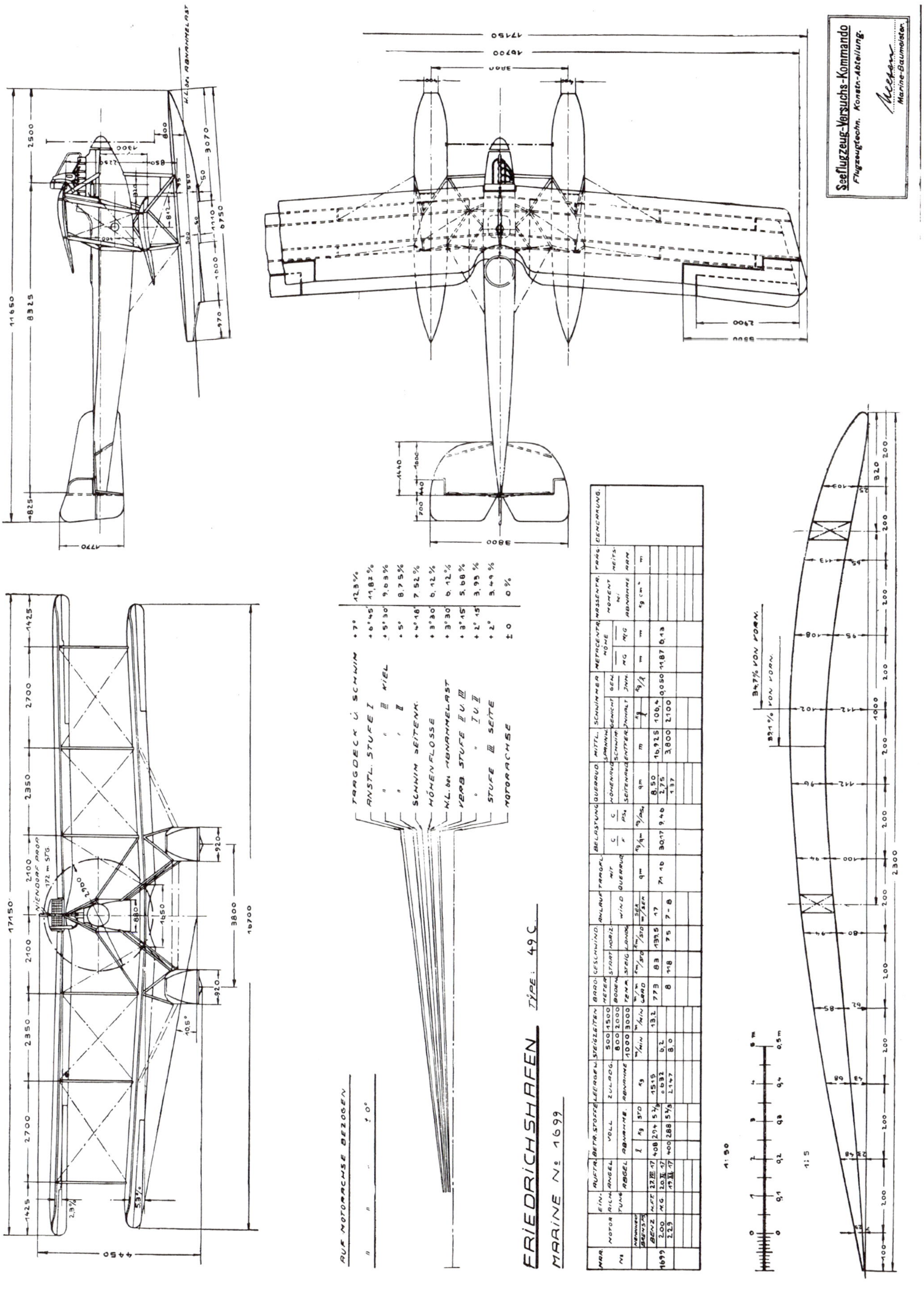
Seeflugzeug-Versuchs-Kommando
Flugzeugtechn. Konstr.-Abteilung.
Marine-Baumeister.
FRIEDRICHSHAFEN
MARINE No 1699
TYPE: 49 C

Friedrichshafen FF59

Above: The FF59 was developed from the FF49 with particular attention to giving the observer the maximum field of fire.

In early 1918 Kober and his team produced the FF59 by modifying FF49C Marine Number 1822. The new machine had larger two-bay wings but a shorter fuselage. To provide the observer with a better field of fire the wire bracing for the inner bay of struts was eliminated and the tail surfaces were re-designed. Two prototypes were built, Marine Numbers 1822 & 1823, but the FF59 was not produced in quantity and the FF49C remained in service until the end of the war.

Friedrichshafen FF59 Specifications

Engine:	200 hp Benz Bz.IV	
Wing:	Span Upper	17.80 m
	Span Lower	16.20 m
	Area	71.5 m²
General:	Length	11.50 m
	Height	4.25 m
	Empty Weight	1,586 kg
	Loaded Weight	2,249 kg
Maximum Speed:		142 km/h
Cruise Speed:		120 km/h
Range:		680 km
Service Ceiling:	800 m	3,500 m

Left: The Zeppelin Musuem at Friedrichshafen has this sample of fuselage fabric from FF49 Marine Number 17XX. In April 1917 the German navy promulgated instructions for a standard camouflage for German seaplanes. The blue-gray color illustrated here was painted on the sides of the aircraft and the upper surfaces were covered with a standard, three-color naval camouflage fabric. Undersurfaces of the fuselage and floats were painted light gray for both improved camouflage and to better protect the aircraft from the elements; however, undersurfaces of the wings were often clear doped finish.

Above: Marine Number 1822 was the first prototype FF59. The vertical tail surfaces were mounted under the tailplane to maximize the observer's field of fire in a manner reminiscent of Brandenburg practice. (The Peter M. Bowers Collection/ The Museum of Flight)

Above: This photo shows Marine Number 1822 after the vertical tail surfaces were modified. Now the fin projects above the fuselage to improve flying qualities but compromising the observer's field of fire.

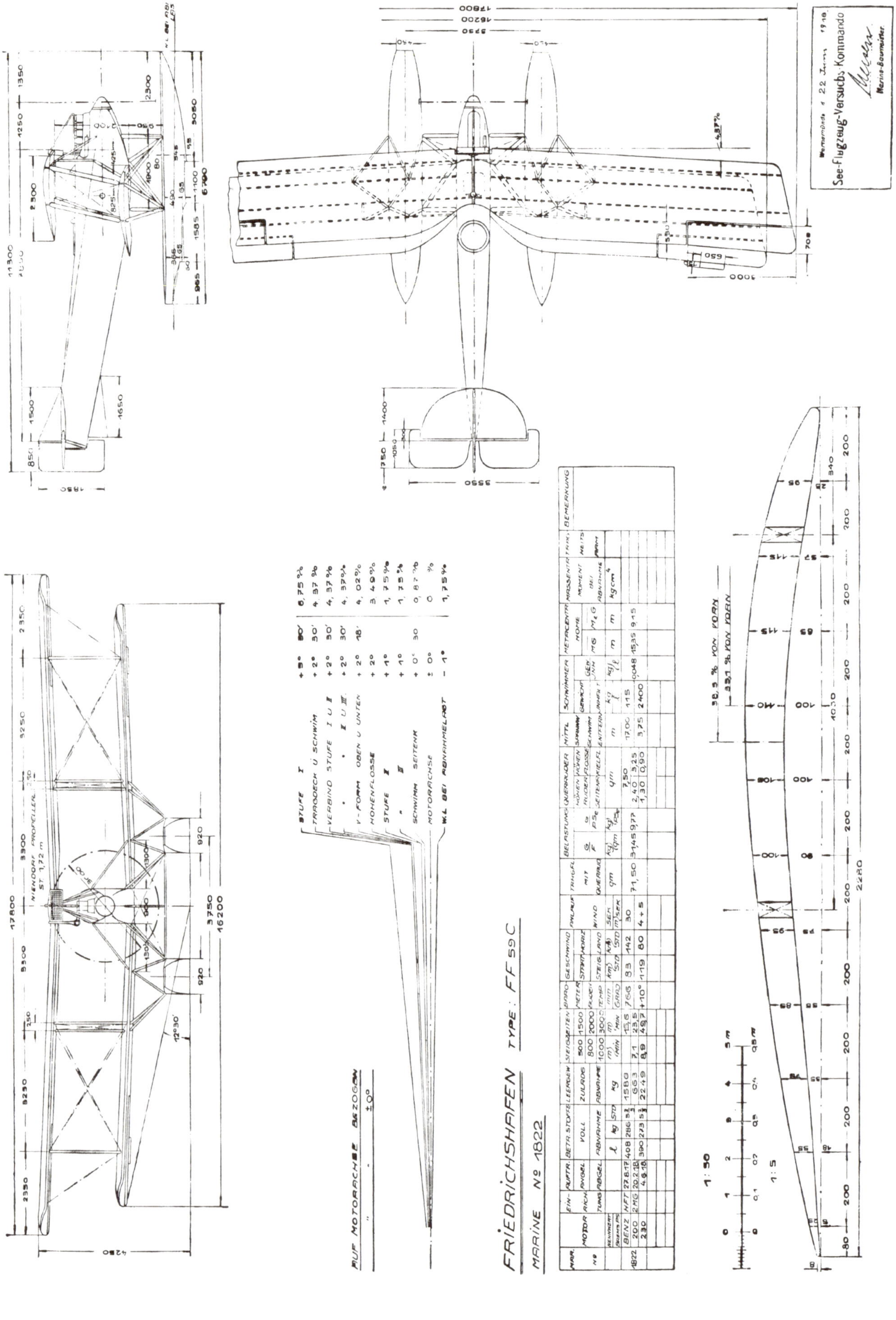

See-Flugzeug-Versuchs-Kommando
Marine-Baumeister
FRIEDRICHSHAFEN TYPE: FF 59C
MARINE Nº 1822

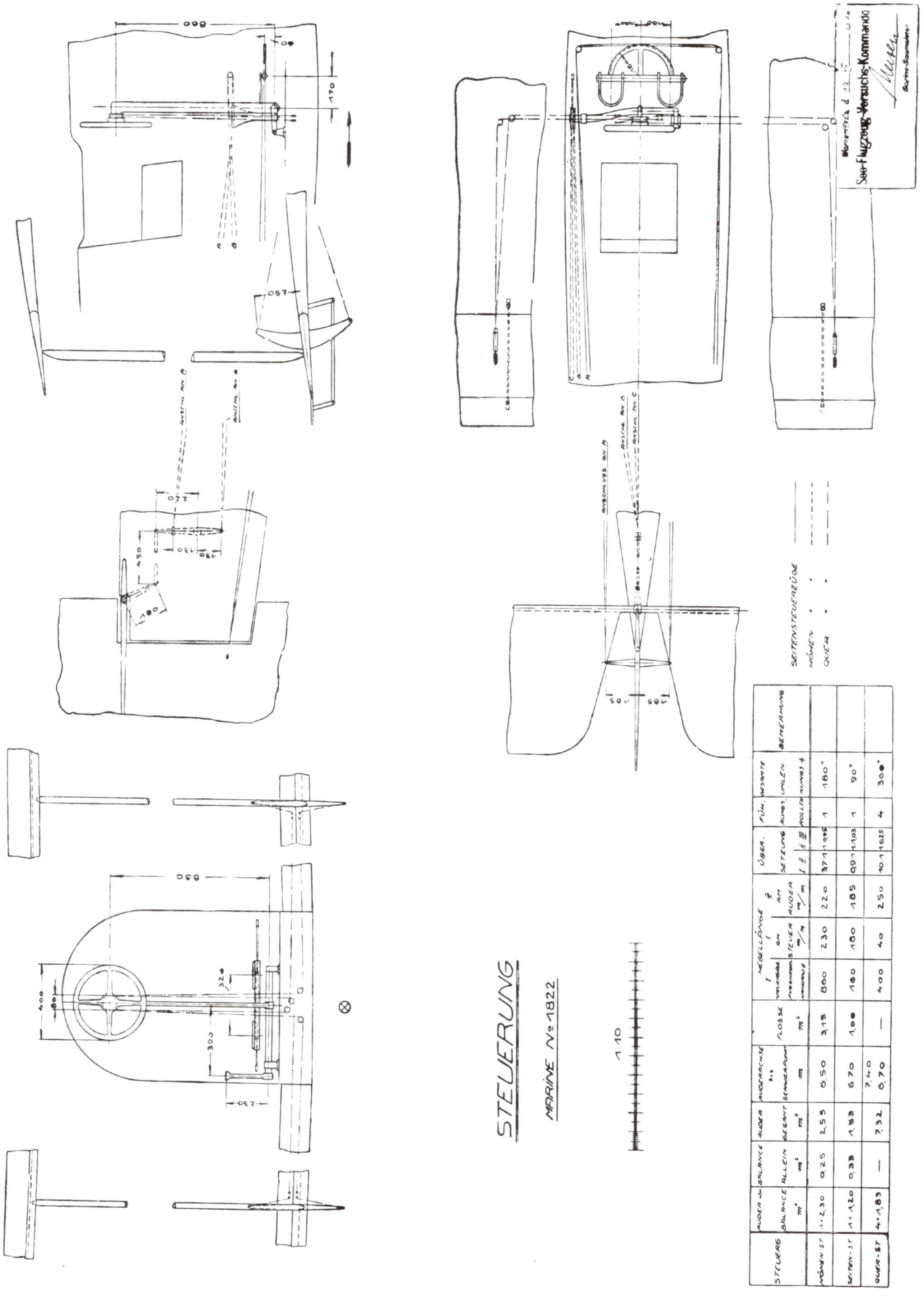

STEUERG	RUDER : BALANCE m²	BALANCE ALLEIN m²	RUDER GESAMT m²	RUDERACHSE bis SCHWERPUNKT m	FLOSSE m²	HEBELLÄNGE I STEUERSÄULE FUSSHEBEL HANDRAD	HEBELLÄNGE I AM STEUER m/m	HEBELLÄNGE II AM RUDER m/m	ÜBERSETZUNG I II III	FÜHRUNGSROLLEN	GESAMTE UMLENKUNG	BEMERKUNG
HÖHEN-ST.	1:2,30	0,25	2,55	6,50	3,15	880	230	220	3,7:1 1,095	1	180°	
SEITEN-ST.	1:1,20	0,33	1,53	6,70	1,00	180	180	185	0,9:1 1,03	1	90°	
QUER-ST.	4:1,83	—	7,32	7,40 6,70	—	400	40	250	1,0:1 1,625	4	360°	

Friedrichshafen FF60

Above: The FF60 during testing on Lake Constance. The FF60 was completed just as the war ended but Friedrichshafen was given permission to test fly the aircraft.

Friedrichshafen's FF60 was the most unusual floatplane the company built, being a large triplane intended to be seaworthy to Sea State 4. The FF60 had four 170 hp Mercedes D.IIIa engines and was designed for long-range maritime reconnaissance. Despite its unusual configuration, its structure was the conventional wire-braced wood covered by fabric construction used by Friedrichshafen for all of its aircraft.

The FF60 carried a crew of four. Defensive armament of this large aircraft consisted of three flexible Parabellum machine guns, and it could carry 1,000 kg of bombs.

Three FF60 aircraft, Marine Numbers 3301–3303, were ordered although as far as is known, only one FF60, likely MN 3301, was finished. Completed just as the war ended, the first flight of the FF60 was in November 1918.

Friedrichshafen FF60 Specifications

Engine:	4x170 hp Mercedes D.IIIa	
Wing:	Span Upper	27.00 m
	Area	235.0 m^2
General:	Height	8.25 m
	Empty Weight	4,875 kg
	Loaded Weight	7,342 kg
Maximum Speed:		140 km/h

Right: Every seaplane had its own *Bordbuch* (logbook) associated with it as illustrated here.

Above: The FF60 during testing on Lake Constance.

Below: The FF60 being launched via crane for testing on Lake Constance.

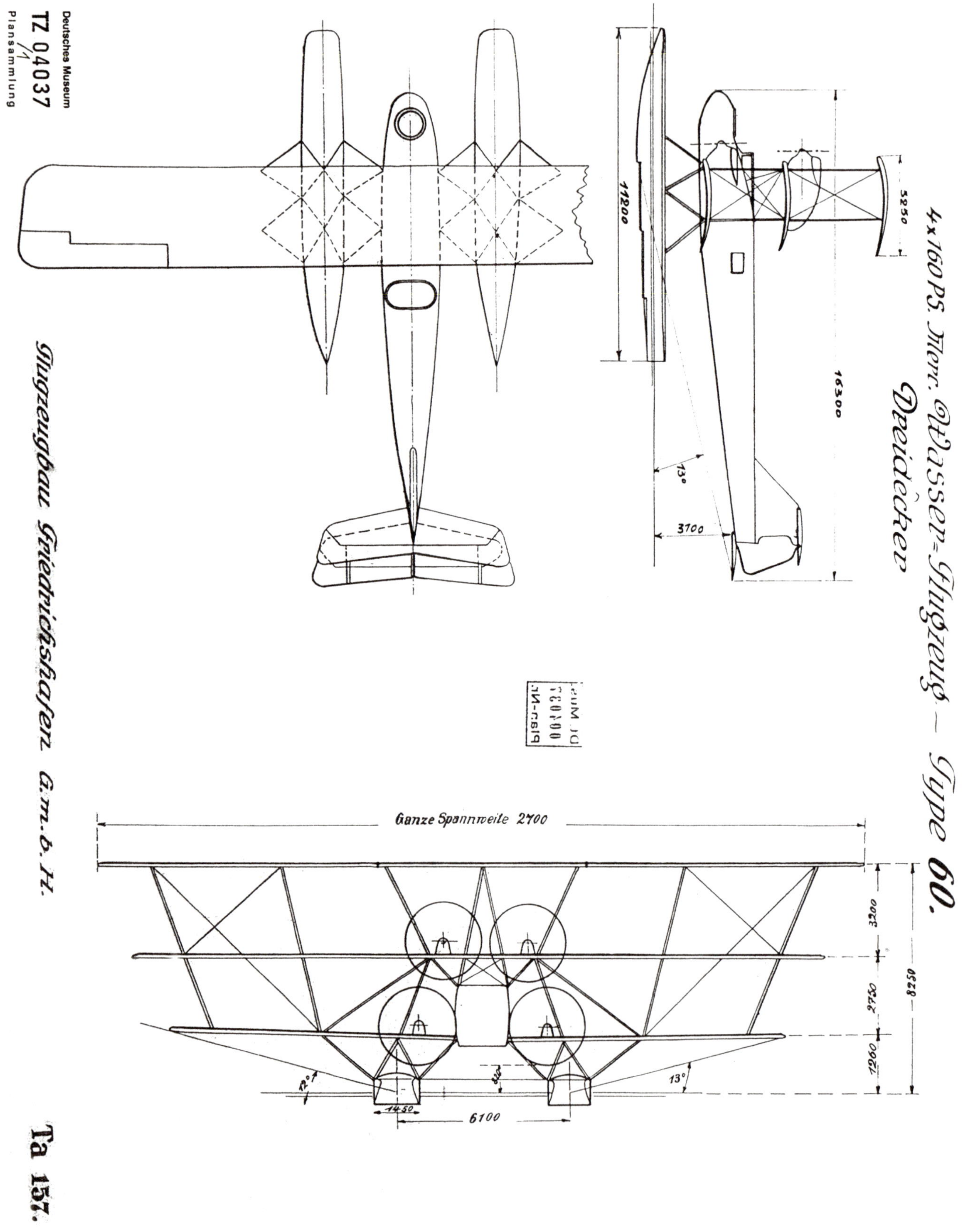

4 x 160 PS. Merc. Wasser-Flugzeug – Type 60.
Dreidecker
Flugzeugbau Friedrichshafen G.m.b.H.
Ta 157.
Deutsches Museum
TZ 04037
Plansammlung
Ganze Spannweite 2700

Above & Below: The FF60 provides a distinctive backdrop for these Friedrichshafen team photographs.

Friedrichshafen FF64

Above: The prototype FF64 front view shows careful construction and streamlining for its time.

After the success of *Wolfchen*, the FF33E carried by the raider *Wolf*, the FF64 was designed as a class CHFT floatplane for use from ships. For this purpose a key design requirement was the ability to fold the wings for easy shipboard storage; another was to be able to launch it quickly.

Three prototypes, Marine Numbers 3061–3064, were ordered in March 1918. SVK acceptance records cease at the end of June 1918, by which time none of the FF64 floatplanes had been accepted. However, the adjacent photos of Marine Number 3061 prove

Friedrichshafen FF64 Specifications	
Engine:	170 hp Mercedes D.IIIa

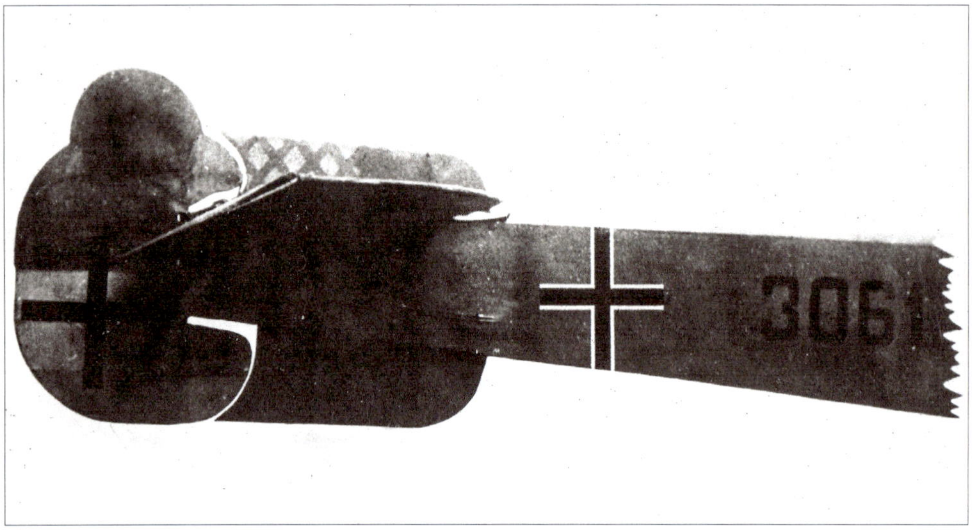

Left: This image of FF64 #3061 front view shows an alternative tail design, with both the rudder and fixed vertical stabilizer surfaces modified from the surfaces shown in the other photos. This is believed to be the third variation tested.

Above: Believed to be the second configuration of the prototype FF64; the tail has an upper fin and the lower fin has been shortened. Marine Number 3061 may have been the only aircraft completed of the three ordered.

that at least one of the three aircraft ordered was completed. The photos also show at least three different configurations for the vertical tail surfaces were built and evaluated. The FF64 was a sturdy, 3-bay biplane able to folding its wings back to a compact footprint for deck handling; additional inner-bay struts enabled bracing wires to be eliminated from that bay for ease of launching.

Above: This is believed to be the first version of the prototype FF64 tail design with no fin on top of the fuselage.

Above: The FF64 was designed to carry a flexible machine gun and wireless to lead a commerce raider to target ships. The tail design seems to have been influenced by Brandenburg designs to give the observer a wide field of fire.

Above: Third version of the (see rudder) FF64 demonstrating its ability to fold its wings for compact shipboard storage. With wings folded the FF64 would fit into a standard cargo hatch.

Above & Below: The FF64 demonstrating its ability to fold its wings for compact shipboard storage, which required additional wing bracing. The ability to quickly launch the aircraft from storage was a key design requirement.

Friedrichshafen Floatplane Fighters

Despite Friedrichshafen's predominance in reconnaissance floatplanes, no Friedrichshafen floatplane fighters reached production. The three Friedrichshafen prototype fighters are described below. They were all competent warplanes but competitor's aircraft were superior.

Friedrichshafen FF43

Above: The prototype Friedrichshafen FF43 on the ramp of the airship hall at the Friedrichshafen works. The assigned Marine Number, #749, has not yet been applied.

Friedrichshafen's FF43, Marine #749, was an original seaplane fighter design powered by a 160 hp Mercedes D.III that first flew on 8 September 1916. It was delivered to the SVK about 30 September but remained a single prototype. Mounting two machine guns, it had a maximum speed of 163 km/h. This was slightly faster than the contemporary Albatros W4 that was selected for production, so the FF43 must have had other shortcomings, and one source notes that it had a poor rate of climb, which is probably why it was not selected for a production order.

Although it remained a single prototype, the German Navy was constantly short of aircraft and the FF43 was assigned to Zeebrugge for operational service. On November 10, 1916, *Ltn.z.S.* Schuler, while flying the FF43, shot down one of two Short seaplanes he encountered during a coastal patrol between Zeebrugge and Dunkirk.

Friedrichshafen FF43 Specifications

Engine:	160 hp Mercedes D.III	
Wing:	Span, Upper	9.92 m
	Span, Lower	9.40 m
	Area	31.0 m²
General:	Length	8.55 m
	Height	3.35 m
	Empty Weight	798 kg
	Loaded Weight	1,078 kg
Maximum Speed:		163 km/h
Range:		400 km
Climb:	1000m	6.0 min
	2000m	9.0 min
	3000m	22.0 min
Service Ceiling:		3,000 m

Above: The prototype FF43 front view shows it was well streamlined for its time.

Above: FF43 outside the airship hall before markings were applied.

Above: The FF43 under evaluation at the *Seeflug-Erprobung* (Seaplane Test Facility) at Warnemünde. With better speed than the Albatros W4 and two guns to the W4's single gun (until later production W4 batches), the FF43 seemed to offer as much potential as the W4. However, the FF43 was not chosen for production, perhaps due to a mediocre rate of climb.

Above: The FF43 prototype with Marine Number and national insignia applied. Despite remaining a single prototype, the FF43 was assigned to Zeebrugge for operational service. On November 10, 1916, *Lt.z.S.* Schuler shot down one of two Short seaplanes he encountered during a coastal patrol between Zeebrugge and Dunkirk while flying the FF43. This was probably Short 184 serial 9054 of Ft.Lt. JD Hume & CPO WE Bradley from Westgate who were KIA in North Sea that day.

Above & Below: The FF43 at the airship hall before its markings were applied.

Above: The FF43 was a well-streamlined aircraft of good proportions that looked a likely prospect for production.

Above: The FF43 without engine cowling shows engine details and the right gun mount.

Above & Below: The FF43 with wings removed shows details of the float struts.

Friedrichshafen FF48

Above & Below: The Friedrichshafen FF48 was a larger aircraft in the class of the W19 and like the W19 was powered by the 240 hp Maybach Mb.IVa. (The Peter M. Bowers Collection/The Museum of Flight)

Friedrichshafen built their FF48 two-seat fighter to the same requirement as the W19. The FF48 had good speed and climb for such a large floatplane but may have lacked the necessary maneuverability, and only three were built, indicating it had no significant advantages over the W19. Certainly it required more wire bracing than the W19 and did not have the look of a fighter. Armament was the standard fixed gun for the pilot and one flexible gun for the observer.

Friedrichshafen FF48 Specifications

Engine:	240 hp Maybach Mb.IVa	
Wing:	Span	16.25 m
	Area	68.0 m^2
General:	Length	11.20 m
	Height	4.40 m
	Empty Weight	1,591 kg
	Loaded Weight	2,216 kg
Maximum Speed:		153 km/h
Cruising Speed:		130 km/h
Range:		740 km
Climb:	1000m	6.4 min
	2000m	14.1 min
	3000m	23.1 min

Friedrichshafen FF63

Above & Below: Powered by the 200 hp Benz Bz.IV, the Friedrichshafen FF63 appears to be a fair design that just does not match the innovative structural design of the W29 and W33 that were its competitors. The additional bracing struts above the wing and the struts above the fuselage added weight and drag that the Brandenburg monoplanes avoided. The tail was made of fabric-covered steel tubing, a major innovation for Friedrichshafen.

Friedrichshafen produced a competitor to the Brandenburg W.29, the FF63. The engine was the 200 hp Benz Bz.IV; virtually no other technical data on the type survives, and even its Marine Number is not known. The FF63 was not as innovative in its structural design as the Brandenburg monoplanes and only a single FF63 was built. The radiator above the engine was surely fitted to expedite early flight testing; this installation obstructed vision and produced far too much drag for a production aircraft.

Friedrichshafen FF63 Specifications	
Engine:	200 hp Benz Bz.IV

Friedrichshafen FF43 SVK Drawing

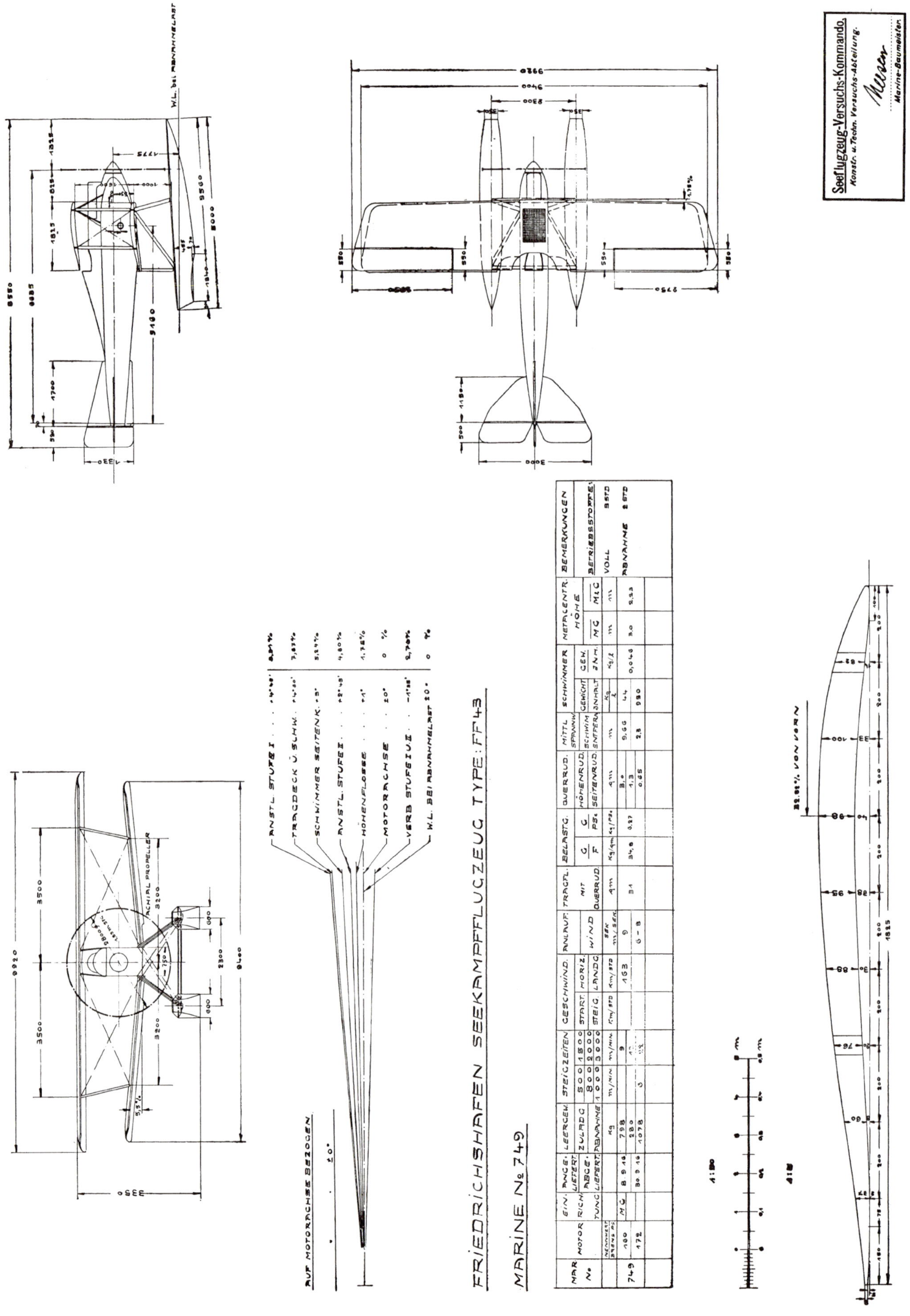

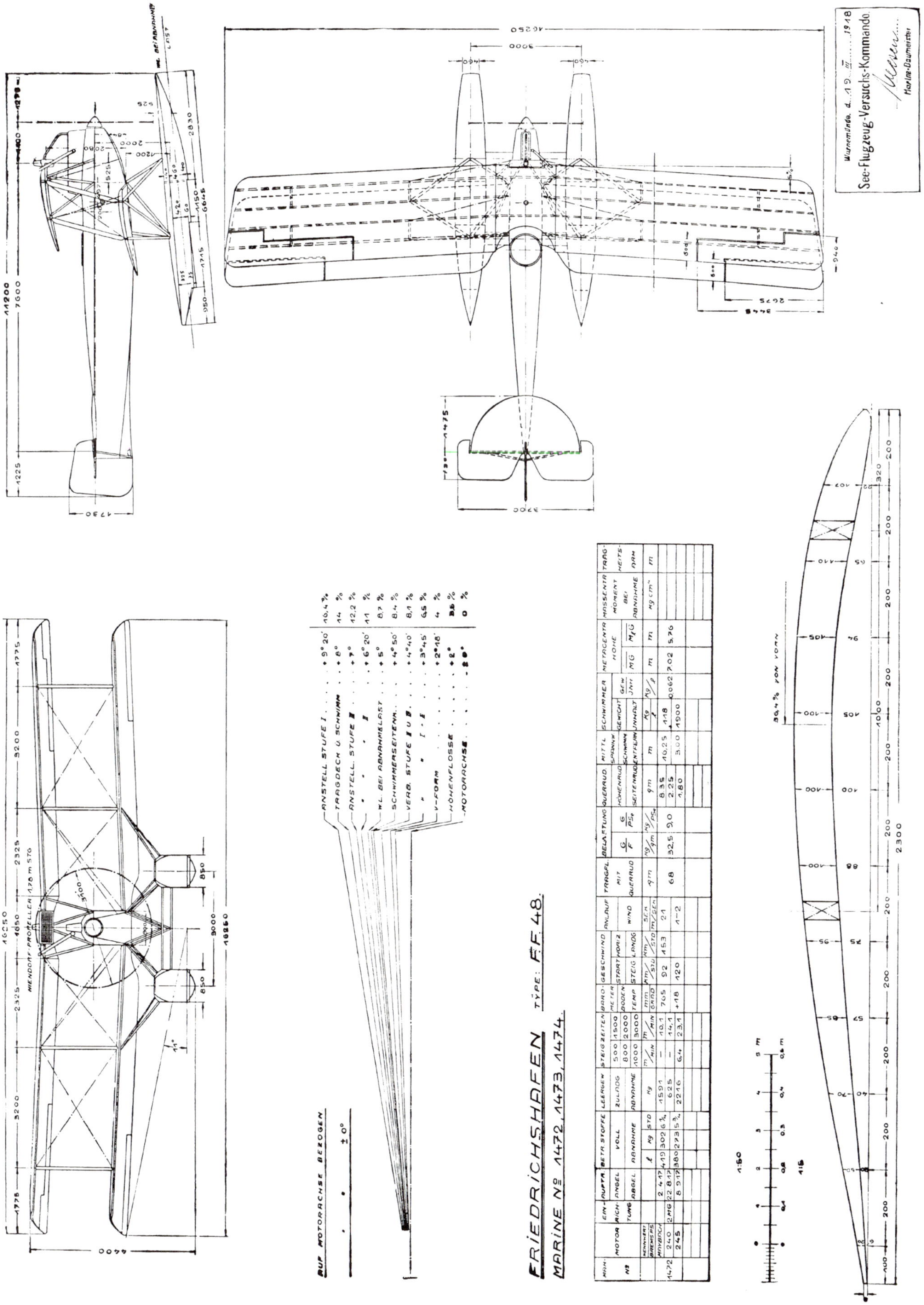
Warnemünde, d. 19. III. 1918
See-Flugzeug-Versuchs-Kommando
Marine-Baumeister
FRIEDRICHSHAFEN TYPE: F.F. 48.
MARINE No 1472, 1473, 1474.
ANSTELL. STUFE I ... +9°20' 16,4 %
TRAGDECK U. SCHWIMM. ... +8° 14 %
ANSTELL. STUFE II ... +7° 12,2 %
" " II ... +6°20' 11 %
W.L. BEI ABNAHMELAST ... +5° 8,7 %
SCHWIMMERSEITENK. ... +4°50' 8,4 %
VERB. STUFE I U II ... +4°40' 8,1 %
" " I-II ... +3°45' 6,5 %
V-FORM ... +2°18' 4 %
HÖHENFLOSSE ... +2° 3,5 %
MOTORACHSE ... ±0° 0 %
AUF MOTORACHSE BEZOGEN
MIENDORF-PROPELLER 1,75 m STG
36,4 % VON VORN
1:50

Friedrichshafen Torpedo Bombers

Above: Front quarter view of FF35 Marine Number 300 shows its basic configuration as a twin-engine floatplane. The 160 hp Mercedes D.III engines are mounted as pushers in primitive nacelles with large, frontal radiators. The horizontal tail was mounted well above the fuselage to keep it out of the water spray from the floats. This was the first Friedrichshafen torpedo plane built; it remained a single prototype.

Before WW1, relatively unsuccessful attempts to drop anti-ship torpedoes from aircraft were carried out in Italy, Great Britain, and the USA. No torpedo bombers were available in Germany, and a quick solution had to be found. The result was land airplanes were converted to carry torpedoes. Both LVG and Albatros built prototypes but tests soon revealed a distinct lack of success.

By the end of 1914 the German Navy defined the requirements for a new torpedo bomber. The crew should be two men and the aircraft should be able to carry a heavy, 700 kg torpedo. Flight endurance must be three to four hours. The German Navy sent development orders to several companies, including Albatros, Brandenburg, Gotha, and Friedrichshafen. As builder of the greatest number of floatplanes for the German Navy, Friedrichshafen was obviously interested in responding to the torpedo floatplane requirement.

Friedrichshafen FF35

Friedrichshafen's initial torpedo bomber design was the FF35 that was ordered in February 1915 handed over to the *SVK* more than a year later, on May 24, 1916, after a maiden flight earlier in 1916. This prolonged development time was due to the factory being fully employed in projects. Powered by a pair of 160 hp Mercedes D.III engines mounted as pushers, the FF35 was a conventional biplane floatplane. The engines were mounted in boxy nacelles with radiators at the front. The aft section was plywood for greater strength and the elevator was raised to protect it from sea spray. The main fittings were forged from steel. The aircraft was designed for a crew of three, a pilot and two gunners, one of whom was also the torpedo aimer. One aircraft, Marine Number 300, was built. Some records indicate a second FF35, Marine Number 310, was built, but official *SVK* records show only the first aircraft being ordered and delivered, with the block of Marine Numbers from 301–400 being left open.

Friedrichshafen Torpedo Bomber Specifications

Specification	FF35	FF41a	FF41at	FF53
Engines	2x160 hp Mercedes D.III	2x150 hp Benz Bz.III	2x150 hp Benz Bz.III	2x260 hp Mercedes D.IVa
Span (Upper), m	23.74	22.00	22.00	—
Span (Lower), m	21.02	20.97	20.97	—
Length, m	13.50	13.70	13.70	—
Height, m	4.50	4.65	4.65	—
Wing Area, m2	100.00	112.50	112.50	—
Empty Wt., kg	2,292	2,300	2,288	—
Loaded Wt., kg	3,543	3,670	3,701	—
Max. Speed, km/h	114	125	121	—
Cruise Speed, km/h	—	115	—	—
Climb to 500 m	—	—	9 minutes	—
Climb to 800 m	—	—	16–18 minutes	—
Climb to 1,000 m	—	—	20–25 minutes	—
Range, km	770	575	605	—
Crew	3	3	2	3
Armament	726 kg torpedo & 1–2 Parabellum guns	726 kg torpedo & 1 Parabellum gun	726 kg torpedo & 1 Parabellum gun	—

Friedrichshafen Torpedo Bomber Production

Type	FF35	FF41a	Unknown*	FF41at	FF53
Quantity	1	1	1	8	3
Marine Numbers	300	678	735	996–1000 & 1208–1210	1663–1665

* This aircraft was powered by two 220 hp Benz Bz.IV engines. It was ordered in May 1916 and the order was cancelled on 18 November 1917.

Friedrichshafen FF41A & FF41AT

The FF41A was an improved development of the earlier FF35. The aircraft was ordered in February 1916 and was delivered on 30 August 1916. The FF41A was powered by a pair of 150 hp Benz Bz.III engines mounted in tractor configuration and had a central fin and three rudders. The Benz engines were more readily available because the Mercedes was in great demand by other high priority aircraft types, especially the Roland C.II at the time of the order, and later Albatros fighters. The engines were mounted in streamlined, sheet-metal nacelles with the radiators mounted above the engines. Compared to the FF35 the wing span was slightly reduced but wing area was increased to support the slightly higher gross weight on slightly less power. Like the earlier FF35, there were three crew positions, but only one machine gun was provided. A single aircraft, Marine Number 678, was built.

Another Friedrichshafen torpedo plane of unspecified type, Marine Number 735 powered by two 220 hp Benz Bz.IV engines, may have been an intermediate step between the FF41A and FF41AT, but with different powerplants. One prototype was ordered in May 1916 but the order was cancelled on 18 November 1917.

The FF41AT was a slightly modified version of the FF41A with a modified fuselage and a single fin and rudder. The engines, dimensions, and armament were the same as the FF41A, but the crew was reduced to two, a pilot and torpedo aimer/front gunner. The first batch of five FF41AT aircraft, Marine Numbers 996–1000, was ordered in October

1916, and the first three aircraft were accepted in May 1917, with the fourth in June and the fifth in July. A second production batch of three aircraft, Marine Numbers 1208–1210, were ordered in January 1917; two from this batch were delivered in July and the third in August.

On the afternoon of August 21, 1917, *Ltn.z.See* Wolfram Eisenlohr was aircraft commander of FF41AT Marine Number 1000 flown by *Flugobermaat* Gruber when they sighted two Russian destroyers at anchor about a nautical mile from the south-eastern coast of Oesel Island in the Gulf of Riga. Eisenlohr directed Gruber to immediately return to their base at Windau to avoid alerting the Russians they had been discovered. Later that day three FF41AT bombers from Windau, each loaded with eight 50 kg bombs in the ventral tunnel designed to hold a torpedo, took off at twilight to attack the destroyers.

The aircraft found the two Russian destroyers at the location where Eisenlohr had seen them earlier, and initiated a bombing attack from an altitude of 600 meters. The destroyers and land batteries immediately commenced heavy anti-aircraft fire. Eisenlohr, the observer and aircraft commander of the second of three aircraft in the flight, observed the first bomber, which dropped four bombs on each destroyer, miss its targets, all the bombs falling short, with the final bomb in the string falling about 10 m short.

Eisenlohr determined to hit his target by dropping all eight of his bombs on one destroyer. He attacked the first destroyer, the *Stroini*, and the fifth bomb in his string of eight scored a direct hit amidships. A plume of fire shot up from the *Stroini*, which sank in shallow water. The third FF41AT missed its target with all bombs. For his victory, Eisenlohr

Above: Front view of the FF35 prototype shows the window in the front of the fuselage for the observer/torpedo aimer.

Above: Side view of the FF35 prototype torpedo bomber on its beaching dolly.

was awarded the Naval Victory Trophy on March 22, 1918. It turned out the *Stroini* had run aground, which is why it was in the same location for the bombing attack.

Eisenlohr also commanded mine-laying missions in the FF41AT. These missions were normally flown at night and the 750 kg (1,654 pound) mines were dropped from an altitude of only 6–8 meters, with a trailing antenna to gauge the altitude. When the tip of the *Schleppantenna* touched the water, a small light went on in the pilot's cockpit to indicate he was at the proper altitude.

Above: Side view of the FF35 prototype torpedo bomber in the water. The German Navy was always short of aircraft, so sometimes used prototypes operationally, and the FF35 was one of those. The FF35 flew limited operations from Libau in May 1916 and flew a number of missions from Angersee over the next four months before being returned to Germany.

Above: Side view of the FF35 prototype torpedo bomber on its beaching dolly outside the assembly hall.

Above & Below: The sole FF41A prototype torpedo bomber afloat outside the assembly hall before its Marine Number and national insignia were applied and engine cowlings were fitted. The multiple rudders are visible.

Above: The FF41A prototype torpedo bomber afloat outside the assembly hall before its Marine Number and national insignia were applied and engine cowlings were fitted.

Above: The tail of this aircraft is not clearly visible in the photo, but it appears to be the FF41A prototype torpedo bomber afloat outside the assembly hall before its Marine Number and national insignia were applied and engine cowlings were fitted.

Above & Below: The sole FF41A prototype torpedo bomber on its beaching dolly outside the assembly hall. The FF41A had a central fin and three rudders; the production FF41AT had a simplified tail with a single fin and rudder.

Above: FF41A Marine Number 678 torpedo bomber with torpedo under test before markings were applied of the engine nacelles were fitted.

Above: FF41A Marine Number 678 torpedo bomber with torpedo under test before the engine nacelles were fitted.

Above: The sole FF41A prototype torpedo bomber on its beaching dolly. The Marine Number was applied in small digits on the nose below the window and the national insignia has been applied to the rudders.

Above: The FF41A Marine Number 678 is in the foreground. At right Brandenburg GW Marine Number 701 is being hoisted into the water carrying a torpedo; another torpedo-carrying GW awaits its turn behind Marine Number 678. Despite being a single prototype, #678 was assigned to Zeebrugge where it was assembled and test flown. Performance was judged to be poor and, as far as is known, no operational missions were flown.

Above: FF41A Marine Number 678 torpedo bomber in the assembly building before the engine nacelles were fitted.

Above: FF41A Marine Number 678 torpedo bomber with torpedo under test before the engine nacelles were fitted.

Above: This aircraft is neither FF41A Marine Number 678 nor a later FF41AT; instead, it may be an intermediate design. Might this be Marine Number 735, the Friedrichshafen torpedo bomber that was cancelled?

Above & Facing Page, Bottom: Marine Number 996, the first FF41AT torpedo bomber built, on its beaching dolly.

Above: Marine Number 997, the second FF41AT torpedo bomber built, is being eased into the water. The production FF41AT aircraft were flown extensively on operations in the Baltic, but these missions were primarily mine laying, conventional bombing, and maritime reconnaissance, not torpedo bombing.

Above: Marine Number 998, the third FF41AT torpedo bomber built, with spherical bombs mounted underneath in the location also used for a torpedo. The aircraft commander's machine gun is protected by a canvas cover.

Above: Marine Number 999, the fourth FF41AT torpedo bomber built. The radiators are mounted above the engines, which are enclosed in streamlined nacelles.

Above: Marine Number 999, the fourth FF41AT torpedo bomber built. (The Peter M. Bowers Collection/The Museum of Flight)

Above: Marine Number 1000, the fifth FF41AT torpedo bomber built and the last aircraft of the first production batch.

Above & Below: Marine Number 1000, the fifth FF41AT torpedo bomber built and the last aircraft of the first production batch, being loaded with a torpedo.

Above: Marine Number 1000, the fifth FF41AT torpedo bomber built and the last aircraft of the first production batch.

Above: Lineup of FF41AT torpedo bombers at Windau; from left, Marine Numbers 1208, 996, and 1000.

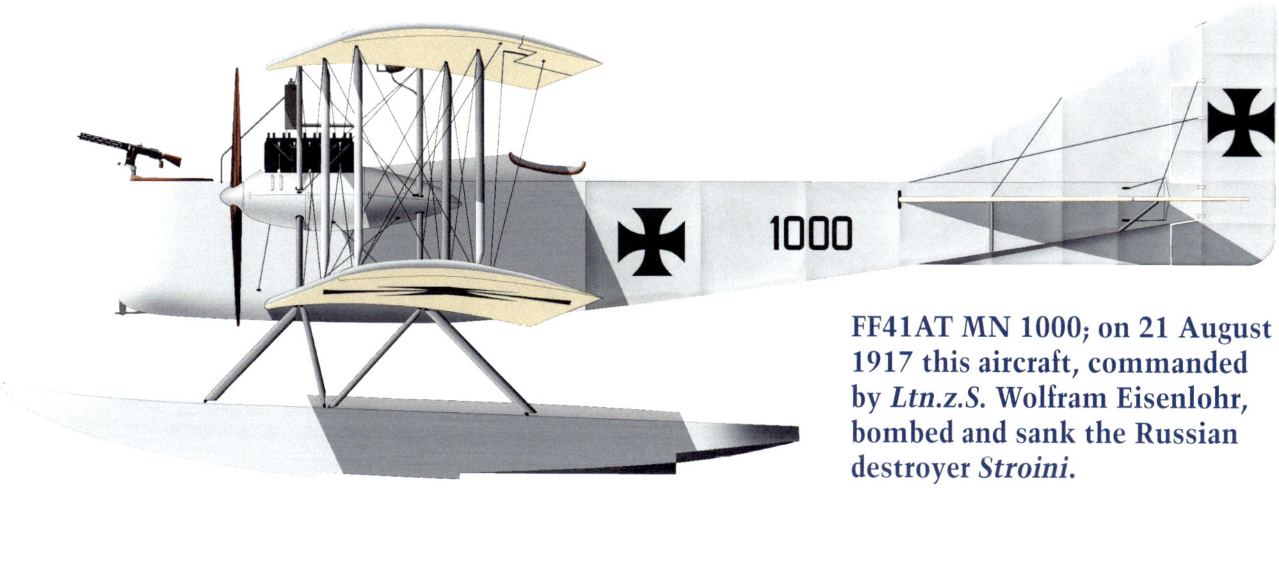

FF41AT MN 1000; on 21 August 1917 this aircraft, commanded by *Ltn.z.S.* Wolfram Eisenlohr, bombed and sank the Russian destroyer *Stroini*.

Above: Launching FF41AT torpedo bombers at Windau; from left, Marine Numbers 996 and 1000.

Above: FF41AT torpedo bomber being inspected by dignitaries.

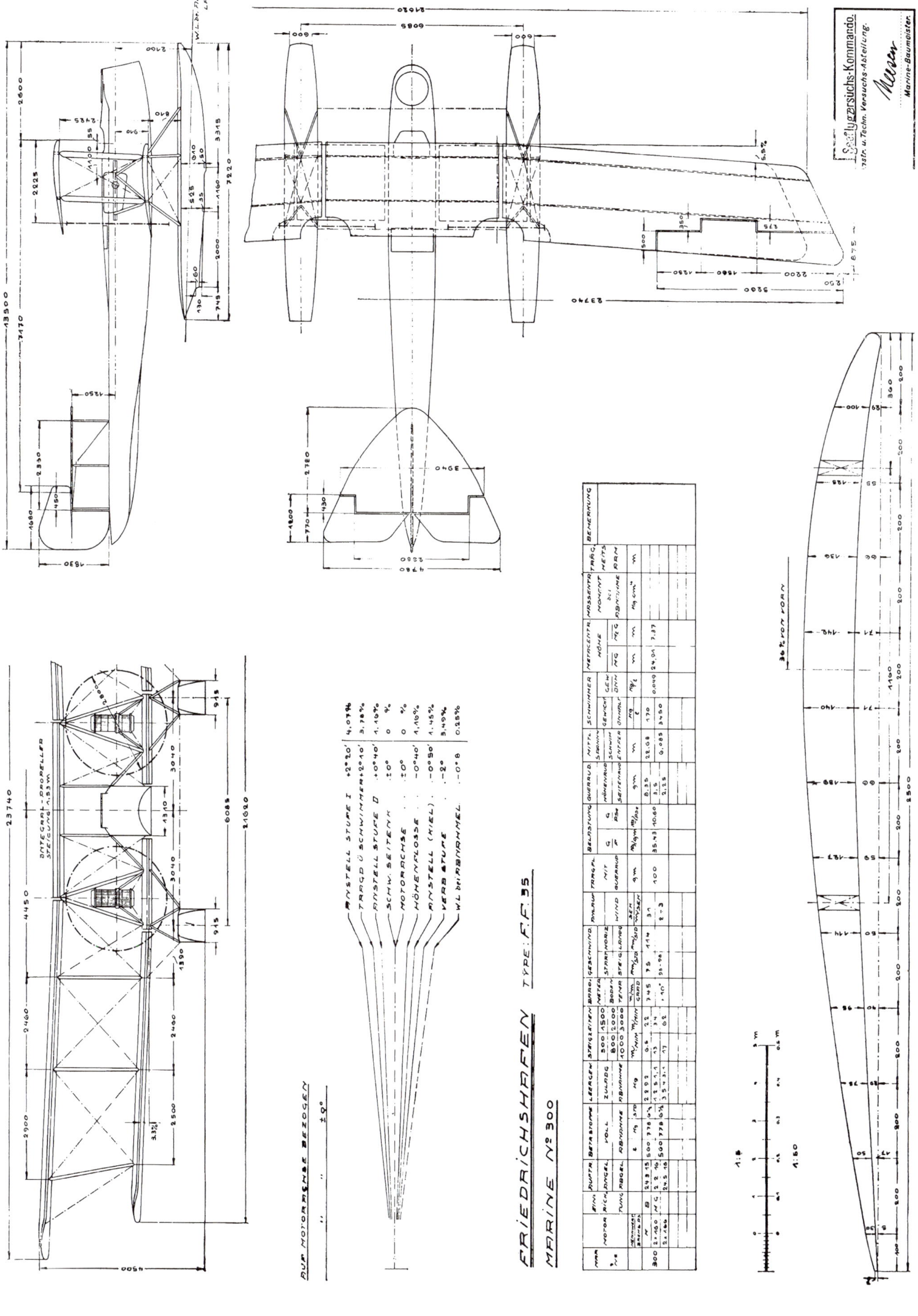
Marine-Baumeister.
FRIEDRICHSHAFEN
TYPE: F.F.35
MARINE No 300
AUF MOTORACHSE BEZOGEN
36% VON VORN

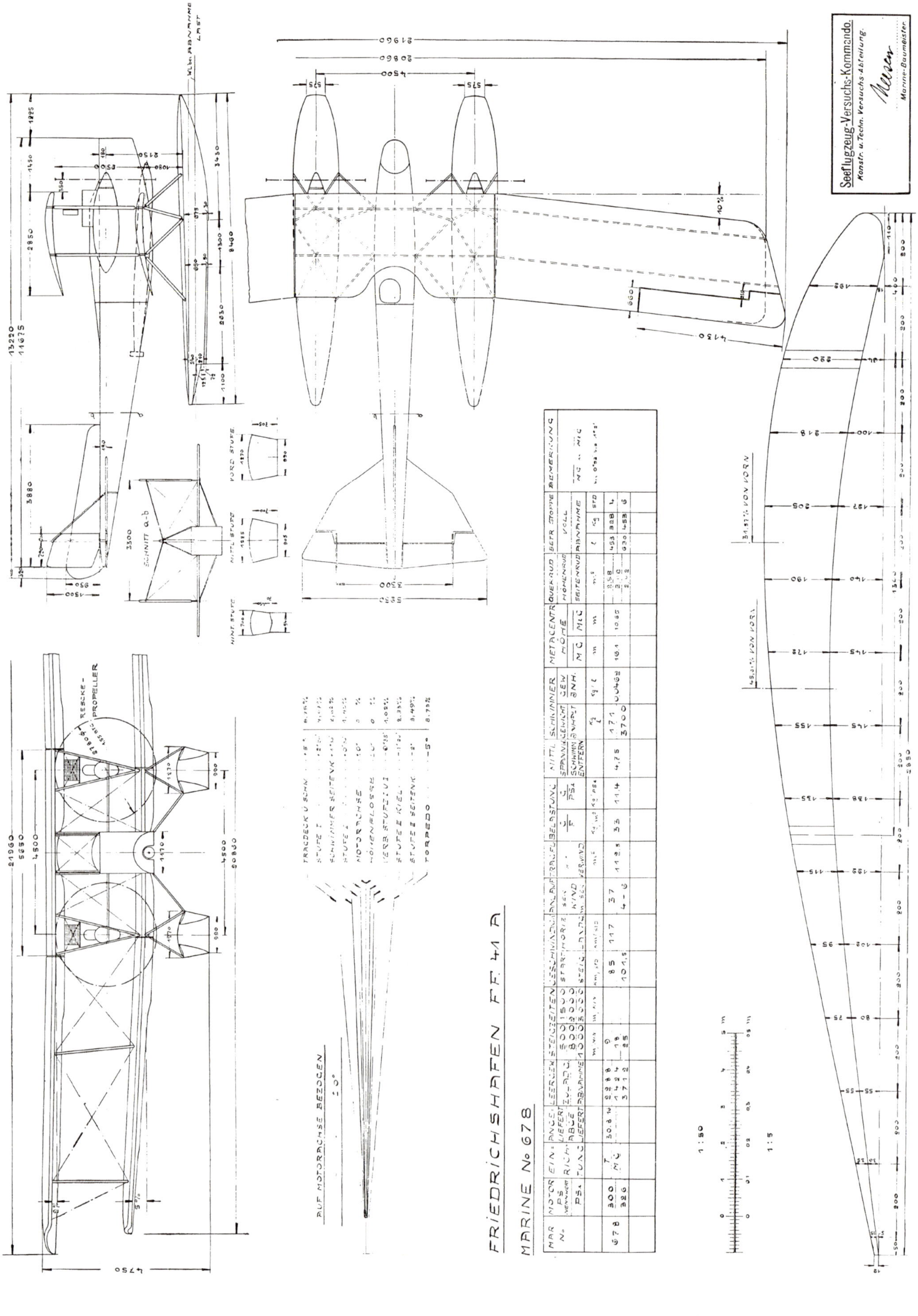
Seeflugzeug-Versuchs-Kommando.
Konstr. u. Techn. Versuchs-Abteilung.
Marine-Baumeister.
FRIEDRICHSHAFEN FF. 41 A
MARINE No 678
RESCKE-PROPELLER
SCHNITT a-b
AUF MOTORACHSE BEZOGEN

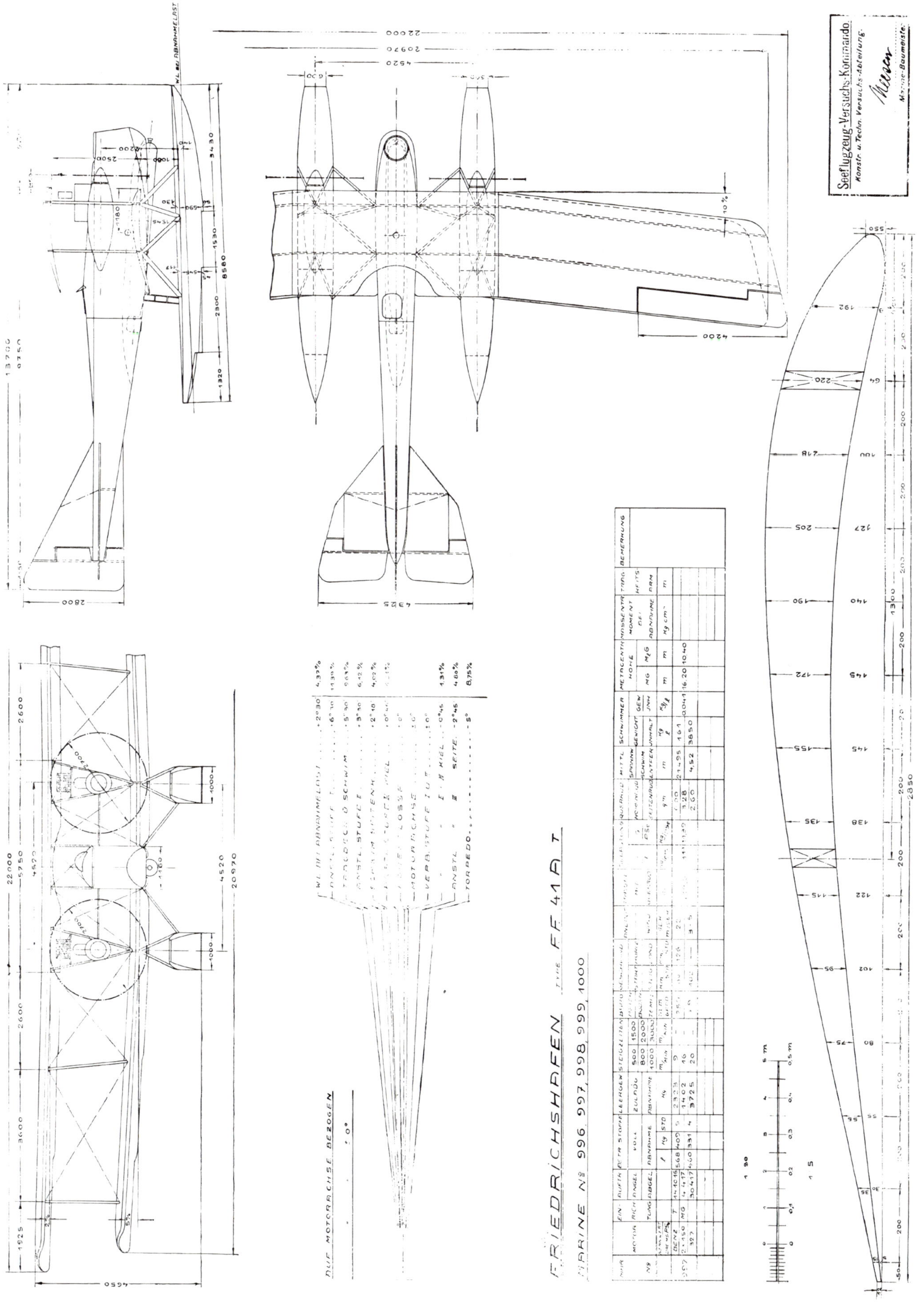
Seeflugzeug-Versuchs-Kommando
Konstr. u. Techn. Versuchs-Abteilung
FRIEDRICHSHAFEN TYPE FF 41 A T
MARINE Nº 996 997, 998 999 1000
AUF MOTORACHSE BEZOGEN

Above, Left & Below: FF41AT torpedo bomber in Estonian service postwar. This aircraft carries number "55" on the fuselage side and an interesting personal marking on the nose. No details of its Estonian service are available. (Courtesy F. Gerdessen)

Friedrichshafen FF53

The final Friedrichshafen torpedo bomber design was the FF53. This was a larger, more powerful aircraft resembling a floatplane derivative of the Friedrichshafen G.IIIa or G.IV land plane bombers. Like the G.IIIa, it was powered by two 260 hp Mercedes D.IVa engines, but unlike the G.IIIa the engines were mounted as tractors to keep the propellers out of the water spray from the floats.

Three aircraft, Marine Numbers 1663–1665, were ordered in June 1917. Photographs verify that at least MN 1663 was completed. Its first flight was on July 30, 1918, and it was handed over to the *SVK* in August. However, *SVK* delivery records after the end of June 1918 are not available, so it is not certain the last two aircraft were completed. By this time the Navy had abandoned torpedo bomber operations so the FF53, now intended for long-range maritime reconnaissance, must have had low priority as the Gotha WD.14 was already fulfilling that role. Ironically, the FF53 finally had the more powerful engines needed to carry a heavier torpedo or bomb load compared to previous Friedrichshafen torpedo bombers, but to no avail. Unfortunately, no specifications for the FF53 have survived.

In Retrospect

Torpedo planes were built in only small numbers. One reason was the higher priority given to army aircraft, and another was the vulnerability of the bombers to anti-aircraft fire during their torpedo run, when they had to maintain a constant course, speed, and altitude before releasing the torpedo. The torpedo bombers were large, slow aircraft flying at extreme low altitude, making them ideal targets for anti-aircraft gunners aboard ship. As a result, torpedo bombing was abandoned after modest successes and most missions flown by these aircraft were reconnaissance, bombing, and mine laying.

Above: Marine Number 1663 was the first of three FF53 torpedo bombers ordered in June 1917. Because SVK records are incomplete after June 1918, the delivery dates for these aircraft are not known. In fact, this photo is the proof the FF53 was actually built and it is not certain the last two aircraft ordered were completed. The FF53 was powered by a pair of 260 hp Mercedes D.IVa engines and greatly resembled the Friedrichshafen G.IIIa land bomber, but the engines were mounted in tractor configuration to keep the propellers out of the spray of water from the floats. Unfortunately, almost nothing is known of the aircraft's specifications or performance and there is no record of its operational use.

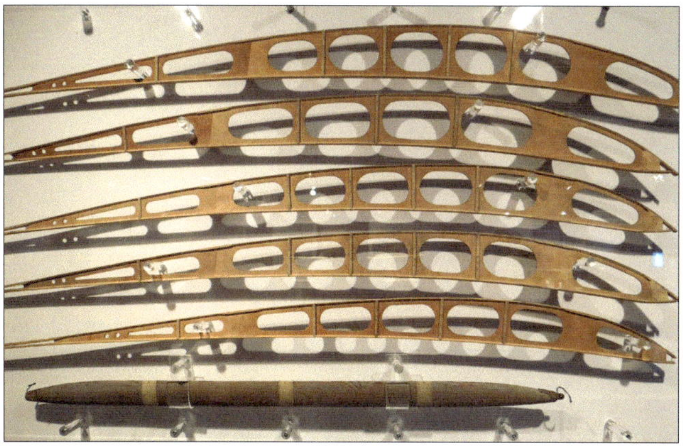

Above: Typical Friedrichshafen wing ribs on display at the Zeppelin Museum.

FF41AT in postwar Estonian service. Serialed '55'; its original Marine Number is not known.

FF43 MN 749 was delivered the end of Sep. 1916. Later assigned to Zeebrugge, on 10 Nov. 1916 *Ltn.z.S.* Schuler downed a Short seaplane while flying it.

FF49C MN 1682 in postwar Netherlands service as V4.

FF33E MN 841 *Wölfchen* in one of the color schemes used during the voyage of *Wolf*; the markings were added after its return to Germany. *Ltn.z.S.* M.A. Stein and *Oberflugmeister* Paul Fabeck, *SMS Wolf*, March 1918.

FF33E MN 841 (ersatz) *Wölfchen* replacement airframe for propaganda after *Wolf's* successful return to Germany. This aircraft may have been an FF33J based on its headrest and nose that looks to have been designed for a spinner.

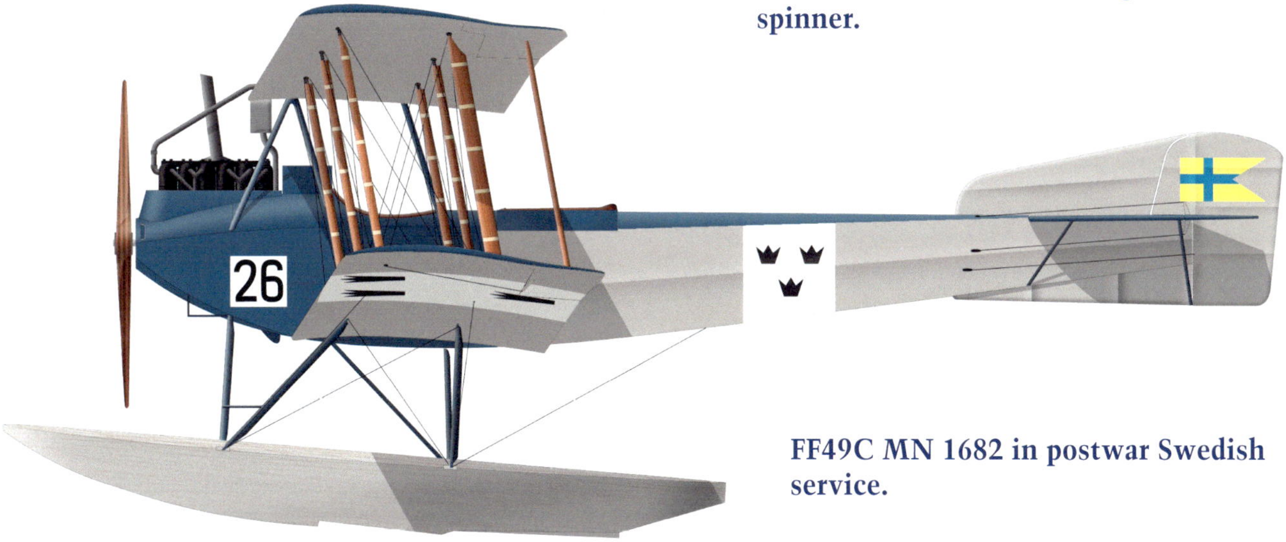

FF49C MN 1682 in postwar Swedish service.

Friedrichshafen Bombers

Although the Friedrichshafen bombers are not nearly as well known as the famous Gotha that bombed London in daylight and later at night, they were actually superior aircraft that served competently as tactical bombers, primarily attacking Allied targets behind the lines at night.

The Friedrichshafen bombers were sturdy, reliable, and perhaps most important to their crews, had good handling qualities that made them much easier to land than the Gothas, which reduced the frequency of their landing accidents with resulting casualties.

Friedrichshafen G.I

Only a single Friedrichshafen G.I, 118/15, was built in 1915. This prototype was powered by two 150 hp Benz Bz.III engines mounted as pushers, and it featured three-bay wing bracing and a biplane tail assembly. Built to the original *Kampfflugzeug* (battle plane) specifications that *Idflieg* issued in 1914, the *Kampfflugzeug* concept was basically an 'aerial cruiser' armed with machine guns and bombs. The aircraft was to have 200 hp, carry a crew of three, and have an endurance of six hours. Unfortunately, the battleplane was a flawed concept.

The Friedrichshafen G.I, factory designation FF36, had a crew of three and a single flexible machine gun in the front gunner's cockpit. Construction of the G.I was the conventional wood frame with wire bracing and fabric covering common for the time. Testing was sufficiently promising that *Idflieg* ordered an improved follow-on design, the G.II.

Above & Facing Page: The Friedrichshafen G.I set the basic configuration for all following Friedrichshafen bombers. Only one prototype was built and evaluated before *Idflieg* ordered an improved, more powerful derivative, the Friedrichshafen G.II. (Peter M. Bowers Collection, Museum of Flight)

Friedrichshafen Bomber Orders				
Date	**Type**	**Qty**	**Serials**	**Notes**
1915	G.I	1	G.118/15	
Late 1915	G.II	6	G.131–136/16	
Feb. 1916	G.II	11	G.100–113/16	G.104/16 was the G.III prototype
Oct. 1916	G.II	12	G.625–636/16	Daimler built. G.637–648/16 not built.
Feb. 1917	G.II	6	G.150–155/17	Daimler built
Nov. 1916	G.III	24	G.1030–1053/16	
May 1917	G.III	25	G.156–180/17	
May 1917	G.III	30	G.240–269/17	Daimler built
June 1917	G.III	100	G.270–369/17	
July 1917	G.III	60	G.373–432/17	Daimler built
Sep. 1917	G.III	6	G.554–559/17	Daimler built
Dec. 1917	G.III / IIIa	60	G.784–843/17	
Dec. 1917	G.III	4	?	Daimler built; serials in range G.433–442/17?
Feb. 1918	G.III	100	G.355–414/18 G.505–544/18	
Feb. 1918	G.III	35	G.150–184/18	Hanseatische built; serials unconfirmed
April 1918	G.III	100	G.645–744/18	
April 1918	G.III	50	G.1400–1449/18	
April 1918	G.III	50	G.1793–1827/18	Ordered from Gotha; not built
April 1918	G.III / IIIa	30	G.750–779/18	Daimler built
June 1918	G.IIIa	50	G.780–829/18	
June 1918	G.IIIa	30	G.880–909/18	Daimler built
July 1918	G.IIIa	150	G.830–879/18 G.1160–1209/18	
July 1918	G.IIIa	45	G.1000–1044/18	Daimler built
July 1918	G.IIIa	40	G.1120–1159/18	Hanseatische built
Sep. 1918	G.IIIb / G.IV	100	G.1465–1564/18	
Oct. 1918	G.IIIa / G.IV	75	G.1045–1119/18	Daimler built
1918	G.V	3	G.900–902/17?	FF62 used the G.V airframe, Mercedes engines

Above: The Friedrichshafen G.I during flight evaluation, which revealed the need for more power. (Peter M. Bowers Collection, Museum of Flight)

Below: The completed Friedrichshafen G.I being rolled out of the factory. (Peter M. Bowers Collection, Museum of Flight)

Friedrichshafen Bomber Specifications							
	G.I	G.II	G.III	G.IIIa	G.IIIa(Daim)	G.IV	G.V
Engines	2x150 hp Benz Bz.III	2x200 hp Benz Bz.IV	2x260 hp Mercedes D.IVa	2x260 hp Mercedes D.IVa	2x260 hp Mercedes D.IVa	2x260 hp Mercedes D.IVa	2x245 hp Maybach Mb.IVa
Span	—	19.75 m	24.0 m	24.0 m	24.05 m	22.60 m	19.20 m
Wing Area	—	75.5 m²	94.26 m²	94.86 m²	—	87.00 m²	76.00 m²
Gap	—	2.10 m	2.11 m	—	—	—	—
Length	—	11.05 m	12.78 m	12.24 m	12.65 m	12.00 m	10.00 m
Height	—	3.41 m	3.54 m	—	3.40 m	—	—
Empty Wt.	—	2,180 kg	2,560 kg	2,700 kg	3,000 kg	2,880 kg	2,410 kg
Loaded Wt.	—	3,180 kg	3,795 kg	4,200 kg	4,500 kg	4,980 kg	4,214 kg
Useful Load	—	1,000 kg	1,235 kg	1,500 kg	1,500 kg	2,100 kg	1,804 kg
Max Speed	—	150 km/h	150 km/h	140 km/h	140 km/h	142 km/h	—
Climb: 1,000 m	—	—	3.5 min.	—	—	—	—
2,000 m	—	—	10.5 min.	—	—	—	—
3,000 m	—	—	20 min.	—	—	—	—
4,000 m	—	—	36 min.	—	—	—	—
5,000 m	—	—	74.5 min.	—	—	—	—

Notes: 1. The G.I & G.V were fitted with one flexible machine gun; the other types had two flexible machine guns.
2. The G.V airframe was also tested with 260 hp Mercedes D.IVa engines.

Below: The Friedrichshafen G.I under construction. When more powerful engines were added to later Friedrichshafen designs, larger, taller rudders were required. (Peter M. Bowers Collection, Museum of Flight)

Above: The Friedrichshafen G.I during construction; the G.I was designed without a rear gunner's position, a feature added to all operational Friedrichshafen bombers. Other than that, the subsequent Friedrichshafen bombers designs generally followed the basic configuration and structure of the G.I. (Peter M. Bowers Collection, Museum of Flight)

Friedrichshafen G.II

After evaluating the Friedrichshafen G.I, *Idflieg* ordered six examples of an improved, more powerful derivative, the Friedrichshafen G.II, factory designation FF38. Following the basic configuration and construction of the G.I, the G.II differed by having two 200 hp Benz Bz.IV engines, more compact two-bay wing bracing, and a conventional monoplane tail, although the first prototype was also tested with a 'box' tail for comparison. The specified bomb load was 150 kg, an indication the *Kampfflugzeug* was evolving into a bomber. Another indication was that a second flexible machine gun was installed for a rear gunner.

These first aircraft had serial numbers G.131–136/15, indicating they were ordered in late 1915. A second batch of a dozen G.II bombers, serials G.100–111/16, was ordered in February 1916. Unfortunately, the prototype was damaged in a crash during testing. Details are not known, but the result was the fuselage and lower wing structure had to be revised before the G.II was approved for production and operational service. The changes were made in May and the type testing was done in July and August, after which the modified prototype was approved for operational service.

However, the first six G.II bombers were apparently retained for a variety of test and experimental programs. For some reason additional modifications to the wings were required, and these were load-tested on October 14. On November 24 a revised tail structure was load tested. Aircraft G.105/16 was accepted on December 1, 1916, an indication that G.II production was completed early in 1917. In addition to the 18 G.II bombers ordered from Friedrichshafen, a like number were ordered from Daimler as the Friedrichshafen G.II(Daim). Only 35 G.II bombers were delivered as G.104/16 became the G.III prototype.

Above: A Friedrichshafen G.II in flight shows its two-bay wing bracing and typical Friedrichshafen configuration. The horizontal struts between the engine and fuselage below the upper longeron indicate either a prototype or a Daimler-built aircraft. (Peter M. Bowers Collection, Museum of Flight)

Below: Friedrichshafen G.II 131/15, the G.II prototype, in the factory. Here it has been modified from its original configuration by a taller fin and balanced rudder. (Peter M. Bowers Collection, Museum of Flight)

Above & Below: Apparently the Friedrichshafen G.II prototype with early vertical tail surfaces. (Peter M. Bowers Collection, Museum of Flight)

Above: Friedrichshafen G.II G.134/15 from the first G.II production batch.

Below: Another view of Friedrichshafen G.II G.134/15 showing the dramatic nose art that was applied. The horizontal engine struts attached to the upper fuselage longeron characterized production Friedrichshafen-built aircraft after structural modifications were applied to the fuselage. Only prototype Friedrichshafen-built aircraft had these struts mounted lower as seen on G.131/15 on the facing page. (Peter M. Bowers Collection, Museum of Flight)

Above: A Friedrichshafen G.II(Daim), characterized by the horizontal struts between the engine and fuselage attaching below the upper longeron. This may be Daimler test pilot Irrek after the successful acceptance flight of the first G.II(Daim) 625.17 on March 3, 1917.

Below: The prototype Friedrichshafen G.II, G.131/15, at the factory. This photo shows the aircraft before modifications to the fuselage structure because the horizontal struts between the engine and fuselage are attached below the upper fuselage longeron. This feature was shared by Daimler-built aircraft but production Friedrichshafen-built aircraft had the struts attached to the upper longerons. (Peter M. Bowers Collection, Museum of Flight)

Above: A Friedrichshafen G.II(Daim) in the drainage ditch on the Daimler airfield at Sindelfingen. The ditch was the source of many complaints from Daimler test pilots. (Peter M. Bowers Collection, Museum of Flight)

Below: An unfinished Friedrichshafen G.IV (note the four-bay wings and box tail) with an SSW fighter in the foreground. The G.IV was basically a G.IIIa with strengthened wings for greater bomb-carrying ability. (Peter M. Bowers Collection, Museum of Flight)

Friedrichshafen G.III / G.IIIa

The Friedrichshafen G.III, factory designation FF45, was an enlarged, more powerful development of the Friedrichshafen G.II. Following the basic configuration and construction of the G.II, the G.III differed by having two 260 hp Mercedes D.IVa engines and larger three-bay wings to enable it to carry a heavier bomb load. The G.III also added a nose wheel under the forward gunner's cockpit to prevent nose-overs on landing. This was important because most German bomber losses were due to bad landings, especially at night. The G.III was under development as early as July 1916, well before the G.II reached the front or was even approved for operational service. As noted above, airframe G.104/16 became the G.III prototype.

In November 1916 *Idflieg* ordered 24 production G.III bombers, serials G.1030–1053/16, with the first becoming the type aircraft, the production standard. This aircraft was flying by March 18, but was not sent to Adlershof for the type test until the designer, Kober, was satisfied with its performance and flying qualities. To improve maneuverability, ailerons were added to the lower wings. G.III 1030/16 was approved for operational service on April 17 providing it successfully performed the three-hour flight, passed the static load tests, and some minor installation and structural defects were fixed. These were easily corrected and the G.III was at the front by June 1917.

The G.III was also produced under license by Daimler as the G.III(Daim), Daimler already having produced the Friedrichshafen G.II under license. *Idflieg* ordered the first Friedrichshafen G.III(Daim) bombers in May 1917. On August 10 Daimler-built bomber G.240/17 passed the three-hour acceptance flight and bomber G.241/17 completed the *Typenprüfung* (Type Test) at Adlershof on August 22. Daimler was required to correct more than 30 minor installation defects and the G.III(Daim) was then approved for operational use. In addition to Daimler, the Friedrichshafen G.III was also ordered from the Caspar Hanseatische Flugzeugwerke, a total of 35 aircraft being produced as the Friedrichshafen G.III(Hansa). The Type Test of this license-built G.III was not completed until July 1918.

Although the Gotha bombers became far more famous due to bombing London by day and later by night, the Friedrichshafen G.III was a better aircraft. Used for tactical bombing, the Friedrichshafen G.III had better flying characteristics than the Gotha, which became dangerously tail-heavy after dropping its bombs. In addition to being more stable in flight, the Friedrichshafen G.III also had a more robust landing gear. Both characteristics became more important as bombers transitioned to night bombing due to improved Allied defenses.

Different engines were tested to improve high-altitude performance. Maybach Mb.IVa engines of

Above: The Friedrichshafen G.III, which followed the basic configuration of all Friedrichshafen bombers, was the main production type. Its configuration was very similar to the Gotha series but the Friedrichshafen had much better handling qualities that made it a safer airplane to fly, especially during landing. This bomber is G.III 360/17.

Friedrichshafen G.III aircraft '6' of *Bogohl* V, *Staffel* 4, summer 1917.

Friedrichshafen G.III 367/17 aircraft '3' of *Bogohl* 8, *Staffel* 25, crewed by *Oblt.* von Gailsheim, *Rittmeister* von Beckmann, *Lt.* von Polnitz, and *Uffz.* Dorr at Maria Alter Aerodrome, summer 1918. The over-painted "26" on the fin may indicate the aircraft was transferred from *Staffel* 26.

260 hp were tried in one G.III, which improved its climb rate to 5,000 m in 42 minutes, a welcome improvement over the standard aircraft. Another G.III(Daim), 245/17, was tested with the over-compressed 260 hp Mercedes D.IVaü engine in January 1918. However, a combination of limited availability of the altitude engines and the switch to night bombing, which was done at lower altitudes, meant the altitude engines were reserved for the high-altitude reconnaissance aircraft that needed them. High reliability was far more important to night bombers than high-altitude performance.

In fact, for night bombing the first priority was reliability of the airframe and engines, followed by increased bomb load. By January 1918 *Kogenluft* wanted the payload increased to 1,000–1,200 kg. Next in priority were good take-off and landing qualities, good flying qualities, maneuverability, and the ability to maintain controlled flight despite an engine failure.

Friedrichshafen addressed these needs with modifications to the basic G.III airframe. A version of the G.III with biplane tail created the G.IIIa. Another derivative with increased wing area became the G.IV, and yet another derivative with tractor propellers became the G.V. Flettner assisted controls were used on the G.IIIa and G.IV, and another G.III was tested with mid-wing ailerons, but these did not go into production.

G.IIIa & G.IIIb

Interestingly, the biplane or 'box' tail had been used on the first Friedrichshafen bomber, the sole G.I prototype. Now a similar tail was tested on the G.III in two versions, with and without a central fin supporting the upper stabilizer. The purpose was to improve directional control after an engine failure. The box tail (*Kastensteuer*) without fixed fin was chosen for production after comparative flight trials were completed in late March 1918. With box tail and a number of additional modifications to improve operational effectiveness, the bomber was designated Friedrichshafen G.IIIa.

At *Idflieg's* urging during a bomber conference that started April 3, 1918, both Friedrichshafen and Daimler agreed to install the box tail on G.III bombers already in production. The additional changes were made when the G.IIIa entered production, the first being delivered by Friedrichshafen in June 1918, from which time the G.IIIa began to replace the G.III. At this same bomber conference it was decided to modify the bomber to enable crewmen to easily move between the forward and aft gun positions. This change

Above: Friedrichshafen G.III 1031/16 of *Kampstaffel* 3.

resulted in the G.IIIb designation, which essentially included all the G.IIIa modifications together with a passageway on the starboard side. The drawings were completed in July and an unknown number of bombers were completed as the G.IIIb.

The G.IIIa proper embodied a number of modifications in addition to the box tail. A new fuel system with increased fuel capacity enabled flight durations up to six hours. A Flettner servo-tab (*Hilfsruder* – assisted rudder) system was fitted to the upper ailerons to reduce the pilot's control forces for easier flight and improved maneuverability. For greater bomb load the standard useful load was increased to 1,500 kg, and the maximum permissible load was increased to 2,100 kg. The landing gear was modified to increase clearance to mount larger bombs under the fuselage. Finally, a Gotha-style gun tunnel was installed in the fuselage to improve the rear gunner's downward field of fire, a change necessitated by increased British night fighter activity. Daimler also changed to G.IIIa(Daim) production as did the Caspar Hanseatische Flugzeugwerke, a total of 40 aircraft being ordered as the Friedrichshafen G.IIIa(Hansa) on July 13, 1918.

Austro-Hungarian Involvement

The Austro-Hungarian *Luftfahrtruppe* (*LFT*) also wanted the Friedrichshafen G.III, but their initial request in November 1917 was refused. In place of the Friedrichshafen G.III the *LFT* was offered the Gotha G.IV and purchased a batch of 40. However, Germany did release one Friedrichshafen G.III airframe to the *LFT* in April or May 1918 for the *LFT* to explore the possibility of license production. Meanwhile the Gotha G.IV(LVG) bombers in *LTF* service were experiencing severe problems that rendered them virtually useless, and the *LFT* again requested 50 Friedrichshafen G.III bombers. This time the request was approved, but a large fire at the Friedrichshafen factory on the night of April 13/14, 1918 precluded fulfilling the order. Due to increased license production by Daimler and Hansa, *Idflieg* finally approved the sale of 20 Friedrichshafen G.IIIa(Daim) bombers to the *LFT* on August 1, but none were delivered. On May 18, 1918 the Austro-Hungarian *Fliegerarsenal* awarded a contract for 100 bombers to the Oesterreichishe Flugzeugfabrik AG (Oeffag), 50 of which were to be the Friedrichshafen G.IIIa, which the *LFT* regarded as the most effective German two-engine bomber at that time. Two Friedrichshafen G.IIIa aircraft with Austro-Hungarian engines installed were sent as pattern aircraft. These aircraft, G.IIIa 789/17 and 790/17, with 250 hp Benz Bz.IV(Mar) engines, were shipped by train and arrived at the Oeffag factory in August. The German drawings had to be revised to Oeffag standards and final assembly of the prototype, Friedrichshafen G.IIIa(Oef) 54.01 began in September. However, with the Austro-Hungarian Empire crumbling, none of these bombers were completed before the Armistice.

Below: Friedrichshafen G.IIIa G.789/18 at Oeffag in August 1918; this aircraft had 250 hp Benz Bz.IV(Mar) engines built in Austria instead of the usual 260 hp Mercedes. The G.IIIa featured a number of improvements over the G.III. The most obvious was the 'box' tail to improve directional control with one engine inoperative. Other improvements included a 'Gotha tunnel' to improve the rear gunner's downward field of fire, greater fuel capacity for longer range, greater bombload, and Flettner servo-tabs on the ailerons to reduce the control forces and thus the pilot's workload.

Left: The passage allowing movement between the rear cockpit and the pilot's cockpit was introduced on the Friedrichshafen G.IIIb. The forward fuselage of this aircraft has not yet been completed.

Below: Friedrichshafen G.IIIa(Daim) 1056/18 was turned over to the French according to the terms of the Armistice. Ordered in October 1918 and in pristine condition, it was probably delivered straight from the factory.

Below: Friedrichshafen G.III 179/17 was covered with the typical dark, night bomber printed camouflage fabric.

Friedrichshafen G.III 180/17 aircraft '4' of *Bogohl* I, *Bosta* 2, downed in Belgian lines in February 1918. It was later painted in Belgian markings, including the Belgian royal coat of arms on the nose. It wears typical night camouflage fabric with two-tone blue mottle on ply and metal surfaces. All crosses were outlined.

Friedrichshafen G.III 369/17 aircraft '5' of *Bogohl* 8, *Bosta* 25, crewed by *Lt.* Reinlein, *Lt.* Wetzlar, and *Vzfw*. Christl at Maria Alter Aerodrome, May 1918.

Friedrichshafen G.III G.1031/16 of *Bogohl* I, *Bosta* 3, flown by *Oblt*. Kupinger and *Vzfw*. Tillmanns during summer 1917. Finish is the original two-tone blue with CDL (clear-doped linen) undersurfaces, here made darker with black squiggles.

Friedrichshafen G.IIIa of *Bogohl* I, *Bosta* 2. The manner in which the crosses were over-painted indicates around June 1918.

Above: Friedrichshafen G.III and G.IIIa bombers of *Bogohl* I in 1918.

Right & Below: Friedrichshafen G.III 180/17 of *Bogohl* I, *Bosta* 2 was brought down near Ghent, Belgium in February 1918.

Above: Friedrichshafen G.III 180/17 of *Bogohl* 1, *Bosta* 2 after capture in February 1918 with Belgian insignia applied.

Above: Friedrichshafen G.IIIa 1429/18 without engines on display in Hyde Park, London on 14 May 1919.

Above: Friedrichshafen G.III being loaded with 50 kg bombs. The radiator details and forward gunner's position are nicely shown. A pintle mount for a second flexible forward-firing machine gun is mounted between the cockpits.

Right: Friedrichshafen G.III 180/17 under evaluation at the French test center at Villacoublay. The nosewheel helped prevent nose-overs during landing, one factor that made it safer to fly than Gothas. Another factor in flight safety was the center of gravity was in the right place, not aft like the Gotha bombers.

Above: Friedrichshafen G.IIIa 862/18 of *Bogohl* 8 in October 1918 during a flight for parachute testing.

Below: Friedrichshafen G.III with aircrew and ground-crew members during bomb loading demonstrating a 300 kg P.u.W. bomb.

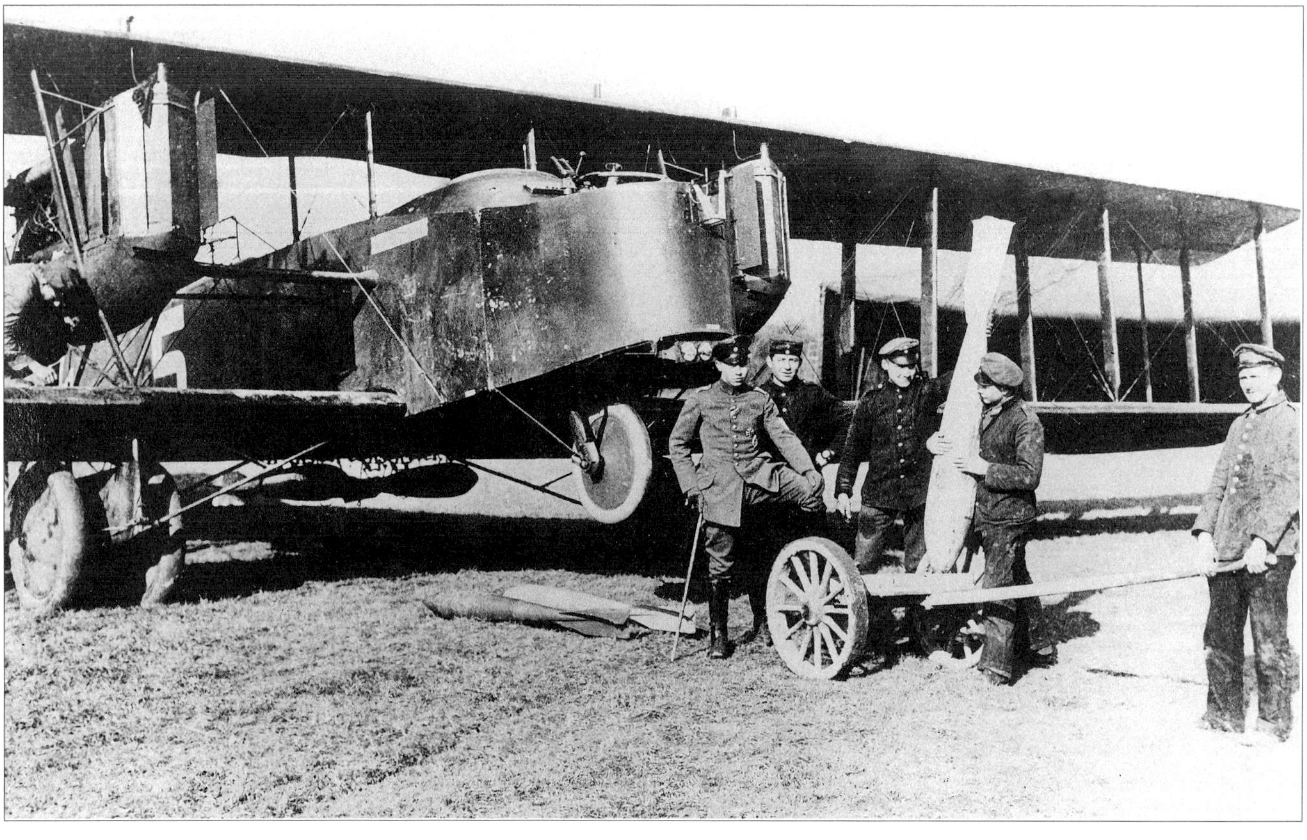

Friedrichshafen G.IIIa 1429/18. When being flown into captivity some German aircraft were ordered to wear a red fuselage band as seen here. This aircraft was photographed on display in Hyde Park, London in May 1919 without engines.

Friedrichshafen G.IIIa No.20 of the Thulin flying school at Ljungbyhed, Sweden in 1920.

Above: Friedrichshafen G.III 1030/16 ordered in November 1916 was the first production G.III.

Below: Friedrichshafen G.III 167/17 in night bomber camouflage. (Peter M. Bowers Collection/Museum of Flight)

Below: Friedrichshafen G.III 180/17 of *Bogohl* 3 brought down in Belgium February 18, 1918, in its original markings.

This Page: Friedrichshafen G.III 180/17 of *Bogohl* 3, *Bosta* 2 brought down in Belgium in its original markings.

This Page: Friedrichshafen G.III 180/17 of *Bogohl* 3, *Bosta* 2 brought down in Belgium in its original markings. It carried a crown insignia on a shield. The bracket for the landing lights is visible under the nose.

Above: Friedrichshafen G.III 180/17 of *Bogohl* 3, *Bosta* 2 brought down in Belgium. The image shows details of the engine, propeller, and rear gunner's cockpit.

Right: Friedrichshafen G.III 180/17 of *Bogohl* 3, *Bosta* 2 brought down in Belgium. The image shows details of the engine cowling and radiator. The aircraft type and serial are clearly visible on the engine cowling.

This Page: Friedrichshafen G.III 180/17 of *Bogohl* 3 brought down in Belgium after conversion to Belgian markings.

Above: This crashed Friedrichshafen G.III shows the hatch in the bottom of the fuselage to enable the rear gunner to fire downward to the rear to defend against fighters.

Above: "Gotha" gun tunnel added to Friedrichshafen G.IIIa.

Above: Becker 2cm cannon tested in Friedrichshafen G.IIIa.

Above: Friedrichshafen G.III(Daim) 241/17 was the second G.III built by Daimler under license. In an attempt to maximize its performance by minimizing drag the pusher propeller even has a spinner fitted.

Above: Friedrichshafen G.III(Daim) 356/17 was one of a batch of 100 ordered in June 1917 and wears a personal marking on the side of its nose. It has no rudder cross and the fuselage cross does not have a white outline.

Above & Above Right: Engine details of Friedrichshafen G.III(Daim) 779/18, the last aircraft of a batch of 30. Some of these aircraft, like 779/18, were transition aircraft with the G.IIIa box tail (*Kastensteuer*) and Flettner tabs (*Hilfsruder*).

Below: Friedrichshafen G.III(Daim) 779/18 showing its box tail (*Kastensteuer*) and Flettner tabs (*Hilfsruder*), attributes of the G.IIIa that were incorporated into a number of G.III aircraft on the production line.

Above & Below: Friedrichshafen G.III(Daim) 392/17 prepares for a flight from the Daimler factory at Sindelfingern. The rudder cross is located at the top of the rudder, a hallmark of Daimler-built G.III bombers. Interestingly, there are no crosses on the upper wing surfaces.

This Page: Friedrichshafen G.III 828/17 landed in the Netherlands and was interned.

Right: Another view of Friedrichshafen G.III 828/17 that landed in the Netherlands and was interned.

Below Right: Crewmen Bierberstein and Nieber pose by the nose of their Friedrichshafen G.III in celebration of an operational milestone.

Below: Close up of the ground crew preparing a Friedrichshafen G.III for a mission.

Below: Friedrichshafen G.III(Daim) 25X/17 sports its iron cross high on its rudder, characteristice of Daimler-built G.IIIs.

Above: A Friedrichshafen G.III 828/17 comes in for a landing.

Below & Bottom: Frontal views of Friedrichshafen G.III bombers.

This Page: Frontal views of Friedrichshafen G.III bombers.

Above: This Friedrichshafen G.III of Kampfstaffel 3 has unusual unit-applied camouflage under its wings.

Above & Above Right: A Friedrichshafen G.III bomber in a hangar in the Netherlands.

Above: This Friedrichshafen G.III has spinners on its propellers and appears to be at the Daimler factory.

Above: This Friedrichshafen G.III of *Staffel* 3/*Bogohl* I, stationed at Tupigny, France, 1918, displays an elaborate coat of arms on the nacelle. Landing lights are also prominently placed under the nose to provide sufficient light for landing in dark or foggy conditions. *Vzfw*. Sturm was the aircraft commander. (Peter M. Bowers Collection/Museum of Flight

Above: A crew loads bombs on their Friedrichshafen G.III.

Right: Friedrichshafen G.III engine details.

Below: A crashed Friedrichshafen G.III spills its bombload.

Left: The cockpit of a Friedrichshafen G.III. There was only one control wheel; there was a pilot but no copilot. The 12 controls on the center and upper right enable the pilot to manage the flow from the fuel tanks.

Below: Friedrichshafen G.III recognition chart provided to German units.

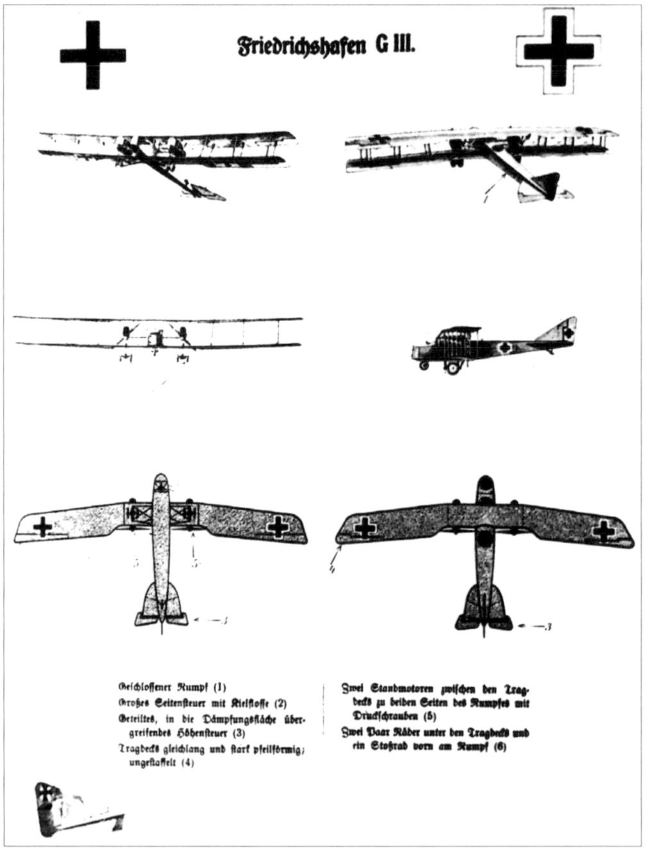

Above Left: Tail details of a Friedrichshafen G.III. The pattern of the night-bomber camouflage fabric shows up well; often it is too dark to show up clearly.

Left: A Friedrichshafen G.III on its take-off run.

Right: The Friedrichshafen G.III was capable of carrying a 1,000 kg *PuW* (*Prüfanstalt und Werft*) bomb as shown here. This bomb was first used in combat in January 1918.

Below: This Friedrichshafen G.III has a test probe fitted on the nose and has rudder stripes indicating an aircraft captured by the Allies. It is likely G.III 180/17.

Below Right: This early model Friedrichshafen G.III still has the windows in the nose like the earlier G.II.

Above: A Friedrichshafen G.IIIb under contruction?

Left: An early-production Friedrichshafen G.III – note the windows in the nose – shares the field with the Albatros D.V D1016/17 of *Ltn.* von der Marwitz of *Jasta* 30 in June 1917.

Left: This view shows the early-production Friedrichshafen G.III is from *Kampfstaffel* 5. The tail of the Albatros D.V of *Ltn.* von der Marwitz is barely visible behind the bomber.

Left: Another view of the early-production Friedrichshafen G.III from Kampfstaffel 5. *Ltn.* von der Marwitz Albatros has been moved. In the foreground *Jasta* 30 pilots prepare for their next combat.

Below: The crew of a Friedrichshafen G.IIIa carrying a 1,000 kg PuW bomb poses with their aircraft.

Above & Below: Friedrichshafen 511/18 was a transition aircraft; it was ordered as a G.III but was given the box tail and Flettner tabs of a G.IIIa during production at the factory.

Below: Friedrichshafen 1425/18 in derelict condition postwar. It was ordered as a G.III but completed with the G.IIIa tail. The fuselage insignia have been cut out for souvenirs. An LVG C.VI missing its rudder rests at left.

Above: The circumstances of this photo are unknown, but the robust Friedrichshafen G.IIIa dwarfs the fragile Bleriot.

Above: This Friedrichshafen G.IIIa has attracted a great deal of attention, probably postwar.

Below: This Friedrichshafen G.IIIa was turned over to the British postwar.

Above: Friedrichshafen G.IIIa tail details.

Above & Below: Friedrichshafen G.IIIa photographed postwar. The box tail and Flettner tabs gave the pilot better control, especially in asymmetric thrust conditions after an engine failed.

Above: Friedrichshafen G.IIIa photographed in British hands postwar.

Above: Photographed in 1920, Friedrichshafen G.IIIa "20" was in Swedish markings at Sweden's Thulin flying school at Ljungbyhed. It had been delivered from Hamburg on 15 January 1920 and here has gained the attention of a group of four men and two women.

Left: Friedrichshafen G.IIIa 1052/18.

Above: Friedrichshafen G.IIIa(Daim) 1037/18 was modified to a passenger transport after the war and registered D 121, is shown here serving with the *Deutsche Luftreederei.* The Treaty of Versailles put an end to this service.

Above: This appears to be the group of four men and two women photographed in Friedrichshafen G.IIIa "20" at Sweden's Thulin flying school at Ljungbyhed with the aircraft in the background.

Above: Friedrichshafen G.IIIa 778/18.

Below: A Friedrichshafen G.IIIa in British markings postwar.

Friedrichshafen G.IV

The Friedrichshafen G.IV, factory designation FF61, was a development of the Friedrichshafen G.IIIa to carry a larger bomb load. The G.IV retained the two 260 hp Mercedes D.IVa engines and box tail of the G.IIIa coupled with stronger four-bay wings. According to one reference 100 were ordered from Friedrichshafen and 35 from Daimler. The Caspar Hanseatische Flugzeugwerke was also to produce the type but apparently none were built before the Armistice. The G.IV was the penultimate development of the Friedrichshafen bomber family, the last to reach operational service, and the last to closely resemble the earlier bombers in the series, especially the G.IIIa.

Above & Below: The Friedrichshafen G.IV was the final Friedrichshafen bomber type to see service. It was a refinement of the G.IIIa designed to carry a heavier bomb load and had four-bay wing bracing.

Above: The only significant difference between the G.IIIa and G.IV was the G.IV had a modified wing with four bays of bracing struts instead of the three bays of the G.IIIa.

Above: Another view of Friedrichshafen G.IV(Daim) 1074/18 also shown on the facing page. This aircraft was from the final batch ordered from Daimler in October 1918 and appears incomplete; perhaps construction was stopped due to the Armistice.

Above: This rear view shows the tail details of the Friedrichshafen FF62, the version of the Friedrichshafen G.V airframe that used Mercedes D.IVa engines. The tractor engine configuration was not used in earlier Friedrichshafen bombers, but elimination of the forward gun turret enabled the engines to be mounted closer to the fuselage centerline if they were mounted as tractors.

Above & Left: Experimental tail surfaces tested for the Friedrichshafen G.IIIa shows the fixed center fin that was tested but did not reach production. The rudders also appear to have longer chord than the production rudders. Designers constantly tried new tail designs to improve flying qualities and controllability after an engine failure.

Friedrichshafen G.IIIa G.823/18 of *Bogohl* VIII, *Staffel* 26, showing the final form of markings.

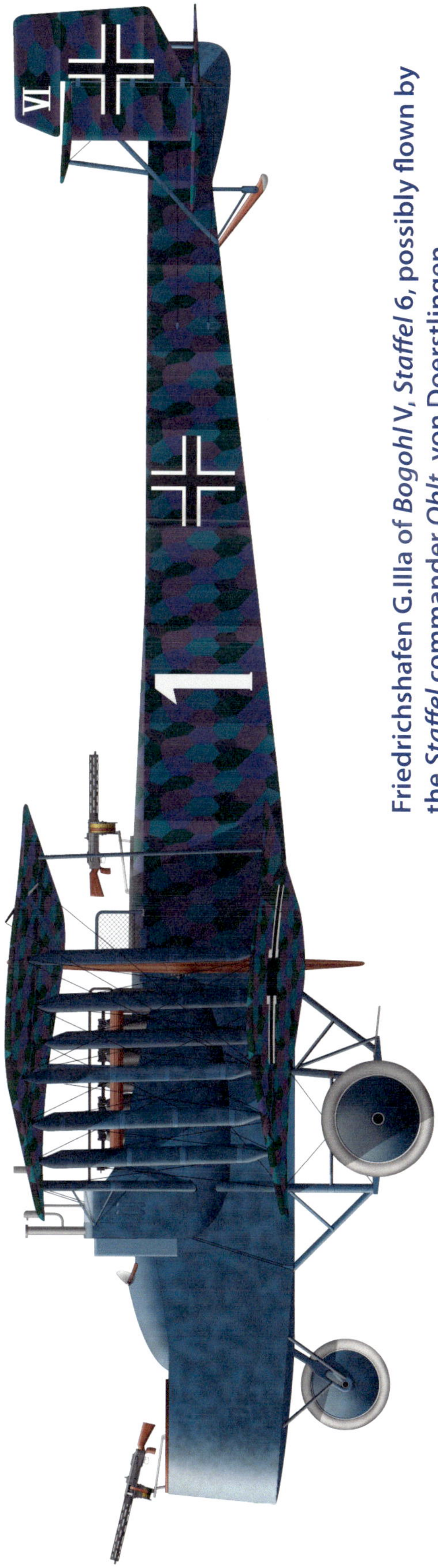

Friedrichshafen G.IIIa of *Bogohl* V, *Staffel* 6, possibly flown by the *Staffel* commander *Oblt.* von Doerstlingen.

Friedrichshafen G.V

The final development of the basic Friedrichshafen bomber family was the G.V. Based on the G.IIIa airframe with box tail using a central fin, the nose turret was removed and the engines were mounted in tractor configuration. Elimination of the nose gun position enabled the engines to be mounted closer together, reducing asymmetric thrust in case of engine failure and making the bomber easier to fly in that situation, which was important for flight safety, especially during night operations.

Two versions were built that differed in the type of engine used, factory designation FF55 being powered by two 245 hp Maybach Mb.IVa engines and factory designation FF62 using two 260 hp Mercedes D.IVa engines. Other than their engines, the two versions appear essentially identical.

Apparently three examples of the FF55 were built and this version was the Friedrichshafen G.V. First flight of the FF55 was on 9 May 1918. The Maybach engine, which developed its rated power at 2,000 m, was more suited to high-altitude performance and was used because the G.V was a contender for high-altitude day bombing, but lost the competition to the faster Gotha GL.VII series. One source states the G.V had provision for a fixed gun for the pilot.

Powered by lower-altitude Mercedes engines, first flight of the FF62 was 20 November 1918. Carrying a heavier bomb load (12x100 kg bombs) than the FF55, the FF62 was almost certainly an attempt to re-purpose the basic design as a night bomber. Apparently only one FF62 was built, and neither version of the G.V saw operational service.

Above: Photographed May 18, 1918, the Friedrichshafen G.V represented a significant modification of the basic Friedrichshafen bomber airframe. The nose turret was removed and the engines were mounted in tractor configuration, enabling them to be moved closer to the aircraft's centerline. This reduced asymmetric thrust after an engine failure, making the aircraft easier to control. Good handling qualities were vital to flight safety, especially during night landings, and the G.V also featured Flettner tabs on the ailerons to reduce the control forces, making flight less tiring for the pilot. The box tail featured a fixed central fin, which was tested but not used on earlier production Friedrichshafen bombers. The Friedrichshafen FF55, given the military designation G.V, was powered by 245 hp Maybach Mb.IVa engines for higher ceiling and speed at altitude with the intention of high-altitude day bombing, although the basic airframe was also tested as the FF62 with the 260 hp Mercedes D.IVa as a night bomber. Only three FF55 aircraft with Maybach engines and one FF62 with Mercedes engines were completed before the Armistice, so the G.V did not see operational service. Photographs showing G.V 900 and G.V 901 are probably G.900/17 and G.901/17. The prominent drag-producing radiators must have adversely affected maximum speed, likely one reason the Gotha GL.VII series was favored for day bombing. The brace supporting the tail has been retouched out of the photo. (Peter M. Bowers Collection, Museum of Flight)

Above & Below: A Friedrichshafen G.V prototype carrying a 1,000 kg P.u.W. bomb photographed on May 9, 1918, the day of its first flight. Doubled wheels are fitted to support the additional weight. Unlike the Friedrichshafen G.IIIa and G.IV, the box tail now has a fixed fin. Like AEG and Gotha, Friedrichshafen continuously experimented the tail configuration of its bombers in a search for improved handling qualities, especially after an engine failure. Provision was apparently made for a fixed gun for the pilot, which makes sense for an aircraft originally intended for high-altitude daylight bombing. The brace supporting the tail has been retouched out of the photos. (Peter M. Bowers Collection, Museum of Flight)

Notes on Friedrichshafen Bomber Serials by Reinhard Zankl

Confirmed Fdh G.IIIb aircraft:	G.1465/18, G.1466/18, G.1470/18, G.1471/18, G.1472/18 (the only confirmed serials in the G.1465–1564/18 range)
Confirmed Fdh G.IV aircraft:	G.508/18, G.526/18, G.827/18, others known in this series are G.IIIa or G.IV
Confirmed Fdh G.IV(Daim) aircraft:	G.1071/18, G.1091/18, G.1093/18, others known in this series are G.IIIa(Daim) or G.IV(Daim)

Miscellaneous Friedrichshafen Land Warplanes

Friedrichshafen was very successful with its floatplanes for the Navy and twin-engine bombers for the Army. Consequently, Friedrichshafen was asked to design a number of single-engine land warplanes for the Army. These were well-built aircraft of conventional design – except for the FF54 multi-plane fighter built during Germany's triplane craze – and like all Friedrichshafen aircraft were of conventional construction. Unfortunately, while well-built and competently designed, Friedrichshafen's single-engine land warplanes were never the best in their category and better-performing designs by other manufacturers were put into production indeed of Friedrichshafen's aircraft.

Friedrichshafen FF37

The Friedrichshafen FF37 was a land-plane version of the earlier FF31 seaplane. Like the earlier FF31, this pusher biplane was a two-seat reconnaissance airplane with the observer in the front cockpit armed with a flexible machine gun.

Unlike the Allies, who persisted with the pusher configuration long after it was obsolete, few German pusher aircraft were built and fewer served operationally. Pushers had unsurpassed forward field of view for their crews, but this did not compensate for their limited performance and vulnerability to fighters. The FF37 was limited to a single prototype by its obsolete configuration and, undoubtedly, its modest performance, although performance specifications have not survived.

Friedrichshafen FF37 Specifications		
Engine:	150 hp Benz Bz.III	
Wing:	Span	16.85 m
General:	Length	9.86 m
	Height	3.70 m

Above: The sole prototype Friedrichshafen FF37; the structure shows clearly through the fabric.

Friedrichshafen C.I

The Friedrichshafen C.I was a land-plane conversion of FF33L seaplane Marine Number 1267. Apparently it was intended as a reconnaissance plane for a Marine unit in Flanders. As would be expected, the observer had a flexible machine gun. The pilot likely did not have a fixed gun in the first version of the aircraft as #1267 was from a batch of FF33L aircraft that lacked a fixed gun for the pilot. The photo is not definitive because a fixed gun, if present, would be on the starboard side of the engine and not visible in any case. Interestingly, a second variant of the Friedrichshafen C.I featured a Mercedes engine without spinner and a fixed gun for the pilot. It is not known if this was modified from #1267 or a second airframe was built.

Friedrichshafen C.I Specifications		
Engine:	150 hp Benz Bz.III	
Wing:	Span, Upper	13.20 m
	Span, Lower	11.70 m
General:	Length	8.0 m
	Height	3.70 m

Above: The prototype Friedrichshafen C.I was converted to a land plane from FF33L Marine Number 1267 and still carries that designation.

Below: A second variant of the Friedrichshafen C.I was powered by a Mercedes engine without spinner and had a fixed gun for the pilot. It is not known if it was modified from Marine Number 1267 or was a second airframe.

Friedrichshafen FF46 (D.I)

Above: The Friedrichshafen FF46, given the military designation D.I, is shown here illustrating a Sanke card. It was a cleanly designed fighter with two machine guns that was developed from the earlier FF43 floatplane fighter.

In September 1916 Friedrichshafen presented a proposal for a single-seat land fighter to *Idflieg*. Developed from the FF43 floatplane fighter, but with no common assemblies, *Idflieg* ordered three prototypes. Power was from the 160 hp Mercedes D.III, the same engine powering the Albatros fighters then arriving at the front, and armament was the two synchronized guns carried by the Albatros, then becoming standard on German fighters. At least two of the three prototypes ordered were built; these differed in their cabane strut design.

Idflieg testing of the Friedrichshafen D.I was prolonged, lasting until April 28, 1917. *Idflieg's* conclusion was that the aircraft's performance and flight characteristics were inferior to the Albatros and series production was not undertaken. Unfortunately, the performance figures that *Idflieg* considered uncompetitive are not available.

Friedrichshafen FF46 (D.I) Specifications

Engine:	160 hp Mercedes D.III	
Wing:	Span	9.00 m
	Area	25.80 m^2
General:	Length	7.10 m
	Height	2.70 m
	Empty Weight	686 kg
	Loaded Weight	901 kg

Right: The logo of the German Integral propeller company.

Above & Below: Friedrichshafen FF46 D2405/16 was apparently the first prototype D.I, identified by its vertical cabane struts. It appears to have been inspired by the Albatros D.I & D.II fighters. Careful attention was given to streamlining for maximum speed, which unfortunately is not known.

Above & Below: This Friedrichshafen FF46 differs slightly from the photos on the previous pages and may be the second prototype D.I, identified by its splayed cabane struts for better field of view for the pilot.

Right: One of the prototype Friedrichshafen D.I fighters from the side.

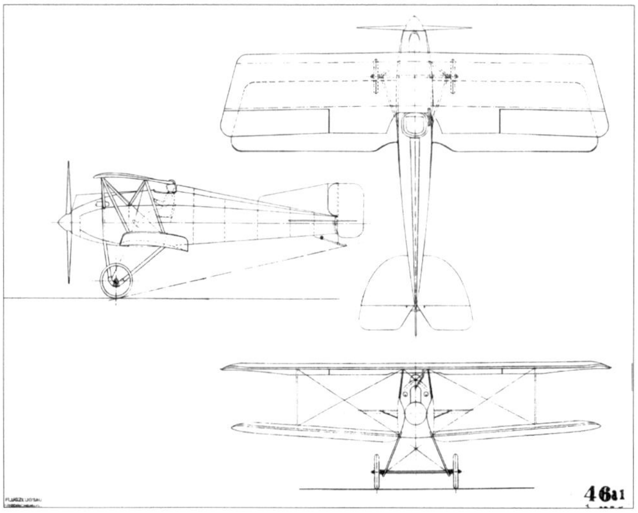

Right: Factory drawing of the first prototype Friedrichshafen D.I fighter.

Below: Friedrichshafen bombers were used as night bombers; this photo shows the landing lights at a typical German bomber field to guide the bombers to a safe night landing. Two rows of these lights, one on each side of the field, were positioned so the returning bomber could land safely between the rows.

Friedrichshafen FF54

The FF54 was Friedrichshafen's response to the Triplane Craze at *Idflieg*. Designed by *Dipl.-Eng.* van Gries, it was a single-seat fighter quadraplane powered by the typical 170 hp Mercedes D.IIIa engine and intended to be armed with the two machine guns standard for German fighters.

The narrow-chord middle wings were mounted to the fuselage with a narrow gap, and the interplane struts between the lower three wings were of broad, airfoil section to reduce drag. The conventional interplane struts between the top wing and upper middle wing were doubtless used to avoid obscuring the pilot's field of view. Flight testing was planned for late November 1917 but the exact date is unknown.

Flight testing soon showed the quadraplane configuration was not successful, and in April 1918 the FF54 was rebuilt as a triplane by removing the lower middle wing and revising the vertical tail surfaces. Renewed flight testing of the FF54 in triplane form began in May 1918, but after a crash in September 1918 the program was terminated. Friedrichshafen was successful designing bombers and floatplanes, but not single-engine landplanes.

Above: The Friedrichshafen FF54 is shown here in its original quadraplane form. No doubt the narrow gap between the pair of middle wings created airflow interference between them and contributed to the aircraft's failure. The lower middle wing was then removed and the vertical tail redesigned for renewed flight testing as a triplane. However, the FF54 was not successful as a triplane either. There is some information that indicates the FF54 was also known as the D.III, but that is not confirmed, nor is there any information on an intermediate Friedrichshafen D.II design. No specifications or performance information on the FF54 has survived.

Facing Page, Top: The Friedrichshafen FF54 is shown here after the lower-middle wing was removed, converting it from a quadraplane to a triplane. (Peter M. Bowers Collection, Museum of Flight)

Friedrichshafen FF54 Specifications	
Engine:	170 hp Mercedes D.IIIa

Facing Page, Bottom: The FF54 after conversion to a triplane in April 1918. The revised rudder design is clearly shown.

Friedrichshafen N.I

Above & Below: Designed to *Idflieg's* single-engine night bomber requirement, the Friedrichshafen N.I was another large, 3-bay biplane like the Albatros C.VIIIN. The pilot was seated well aft with an engine and long length of nose that obscured the view while landing, a feature hardly suitable for safe night operation. The required bomb load is under the wings.

Friedrichshafen specialized in floatplanes and bombers, so seemed well-positioned to develop a suitable N-type, a single-engine night bomber. However, powered by a 260 hp Mercedes D.IVa, the Friedrichshafen N.I appears overly large and cumbersome with its long swept wings and very long nose that would have obscured the pilot's vision during night landings, hardly desirable for its intended night bomber role. In addition, its engine was in demand for other operational types already in production, and it remained a prototype. The AEG N.I selected for production was derived from an aircraft already in production and service and used a smaller engine that was more readily available.

Friedrichshafen N.I Specifications	
Engine:	260 hp Mercedes D.IVa

Friedrichshafen Postwar Floatplanes

After the Armistice Friedrichshafen attempted to continue building aircraft by creating passenger versions of their floatplanes. This did not last long as the Allies were determined to destroy Germany's aircraft industry, and like other German aviation manufacturers Friedrichshafen went out of business.

Friedrichshafen FF67

The Friedrichshafen FF67 was a postwar transport aircraft based on the FF49C. First flight was on 12 December 1918, barely a month after the Armistice. In March 1919 the aircraft was modified by addition of a cabin for the passengers, but the pilot continued to sit in an open cockpit. Power for the single prototype was a 260 hp Mercedes D.IVa.

Above: The FF67 after the passenger cabin had been added. (Peter M. Bowers Collection, Museum of Flight)

Friedrichshafen FF67 Specifications	
Engine:	260 hp Mercedes D.IVa

Right: The FF67 after the passenger cabin had been added.

Friedrichshafen FF71 & FF71a

The Friedrichshafen FF71 was Friedrichshafen's second attempt to create a postwar transport aircraft based on the FF49C. At least one prototype was completed, and its first flight was in May 1919. The FF71 was a large, 3-bay biplane derived from the FF49C. Originally, the aircraft had open cockpits but it was later modified by addition of a cabin for the passengers, although the pilot continued to sit in an open cockpit. Unfortunately, specifications have not survived.

From 1921 it is alleged that LFG built five FF71a aircraft, and the specifications refer to the FF71a.

Friedrichshafen FF71a Specifications		
Engine:	200 hp Benz Bz.IV	
Wing:	Span Upper	17.15 m
	Span Lower	17.15 m
	Area	71.16 m^2
General:	Length	11.65 m
	Height	4.25 m
Maximum Speed:		140 km/h
Cruising Speed:		130 km/h

Above: The Friedrichshafen FF71 transport floatplane as originally built with open cockpits. It still carries the camouflage fabric used on warplanes.

Above: The Friedrichshafen FF71 transport floatplane after modification to have a passenger cabin.

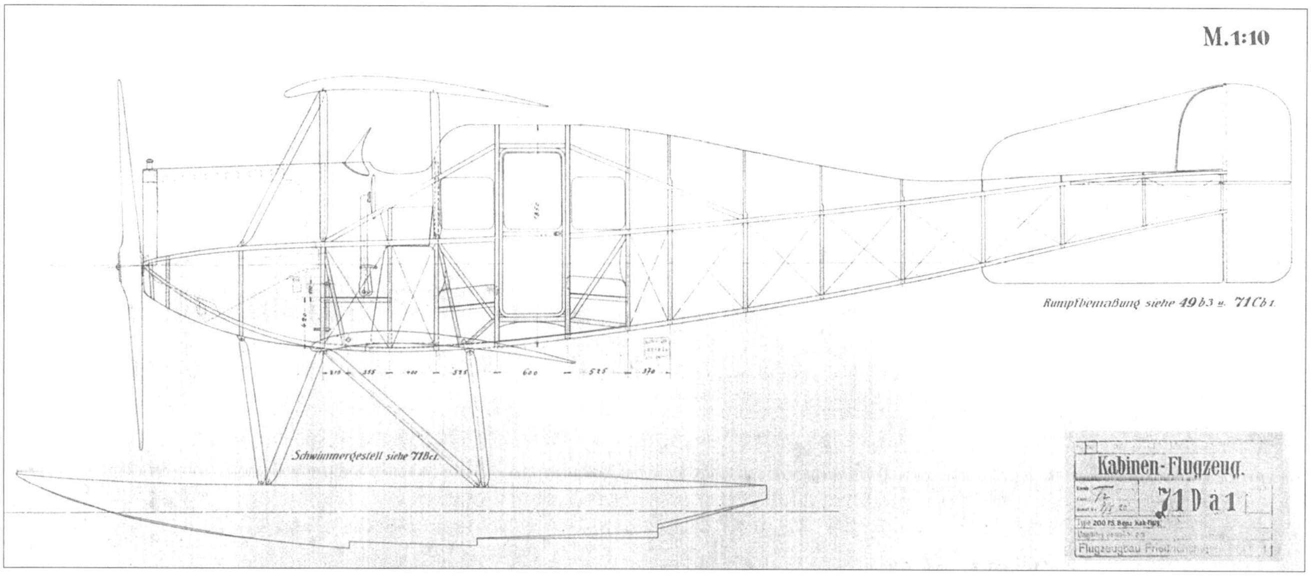

Above: Friedrichshafen factory drawing showing the FF71a civil transport floatplane with cabin was based on the FF49C.

Right: Pilot Truckenbrodt in the cockpit of the Friedrichshafen FF15 before the war.

Above: Prototype Friedrichshafen FF63 two-seat fighter.

Above: Prototype G.II with initial tail design.

Above: Prototype G.II with production tail design.

Above: Friedrichshafen G.IV.

Left: FF62, the prototype G.V with Mercedes engines.

Friedrichshafen in Retrospect

Flugzeugbau Friedrichshafen GmbH was one of the most important WWI aircraft manufacturers in Germany and became the largest supplier of seaplanes to the German Navy, contributing between 36%–42% of all floatplanes (depending on the source) and in also produced a successful series of twin-engine Army bombers.

Friedrichshafen's floatplanes were robust and reliable, essential assets for aircraft flown over water, and were able to operate in higher sea states than most other German floatplanes. These critical qualities were the key to Friedrichshafen's success with naval aircraft.

Similarly, Friedrichshafen's bombers were robust aircraft with good handling qualities. Friedrichshafen bombers were used for short and medium range night bombing missions against Allied targets behind the lines, but because they were not used to bomb England they never gained the fame – or notoriety? – of the contemporary Gotha bombers despite their superior qualities. In particular, the Friedrichshafen bombers did not suffer the aft center-of-gravity problems that afflicted Gotha and Rumpler bombers and, coupled with the nose wheel that prevented over-turning on landing, were therefore were much safer to land at night than the Gothas. In fact, a batch of 50 Friedrichshafen bombers were ordered from Gotha in April 1918, clearly indicating that the Friedrichshafen design was superior, but this batch was not built. However, Daimler and Hanseatische did build Friedrichshafen bombers under license.

Friedrichshafen built a few single-engine fighter, reconnaissance, and night-bomber landplane prototypes for the Army but none of these were placed in production. Friedrichshafen's expertise was in twin-engine bombers and seaplanes; their robust structures were a little heavy to be fully competitive in the area of single-engine landplanes. Although Friedrichshafen built a quadraplane fighter prototype during Germany's triplane craze, its performance was not satisfactory. It was modified into a triplane by removal of one wing, but its performance was still not sufficient to merit a production order.

In addition to this multi-plane experiment, Friedrichshafen also built a large, four-engine triplane floatplane, the FF60, for long-range reconnaissance. The FF60 also remained a prototype, although that may have been due to the Armistice and not unacceptable performance and flying qualities.

Essentially all Friedrichshafen designs adhered to the conventional wood, wire, and fabric structural technology of the time. The late-war prototype FF63 did feature fabric-covered steel tube construction for its empennage, which was a major structural innovation for Friedrichshafen. Moreover, the FF66 design was intended to have metal construction and parts were made and tested, but the aircraft itself was not built due to the Armistice. Friedrichshafen founder Theodor Kober wanted to build all-metal aircraft after the war, but this was not supported because Dornier, working for Zeppelin-Lindau, was already building all-metal aircraft and the Zeppelin concern would not support a second designer creating metal aircraft.

Neither did Friedrichshafen innovate in aerodynamics; nearly all their wartime designs were conventional biplanes. Robust designs and solid manufacturing quality were the secrets to Friedrichshafen's success during the war. As the Allies intended, the Versailles Treaty restrictions drove Friedrichshafen out of aviation postwar.

Left: *Unter-Lt.* Martin Lange by the nose of his FF33L.

Bibliography

Books

Anderson, L & Sanger, R. *Retribution and Recovery*, Air Britain, UK, 2014.

Borzutzki, Siegfried, *Flugzeugbau Friedrichshafen GmbH*, Verlag Markus Burbach, 1993.

Gray, Peter, and Thetford, Owen, *German Aircraft of the First World War*, second revised edition, New York: Doubleday & Company, Inc., 1970.

Green, William, and Swanborough, Gordon, *The Complete Book of Fighters*, Smithmark Publishers, 1994.

Grosz, Peter M., *Fdh G.III–IIIa*, Albatros Publications, Berkhamstead, 1997.

Grosz, Peter M., *FF33E!*, Albatros Publications, Berkhamstead, 1999.

Herris, Jack, *Gotha Aircraft of WWI*, Aeronaut Books, 2013.

Keskinen, K, Partonen, K and Stenman, K. *Suomen Ilmavoimat 1918–27 Finnish Air Force.* K. Stenman, Finland, 2005.

Articles

Grosz, Peter M., "The Cruise of the Wölfchen" *Cross & Cockade* Vol.14, No.1, Spring 1973, p.1–26.

Grosz, Peter M., "Frontbestand" *WW1 Aero* No.107, Dec. 1985, p.60–66.

Grosz, Peter M., "Frontbestand" *WW1 Aero* No.108, Feb. 1986, p.66–69.

Sanz, M. "En Friedrichshafen FF49C i Kalmar," *Sevensk Flyghistorisk Tidskrift* 4/11, P.9.

Internet

http://en.wikipedia.org/wiki/Flugzeugbau_Friedrichshafen

Selected Index

FF33E

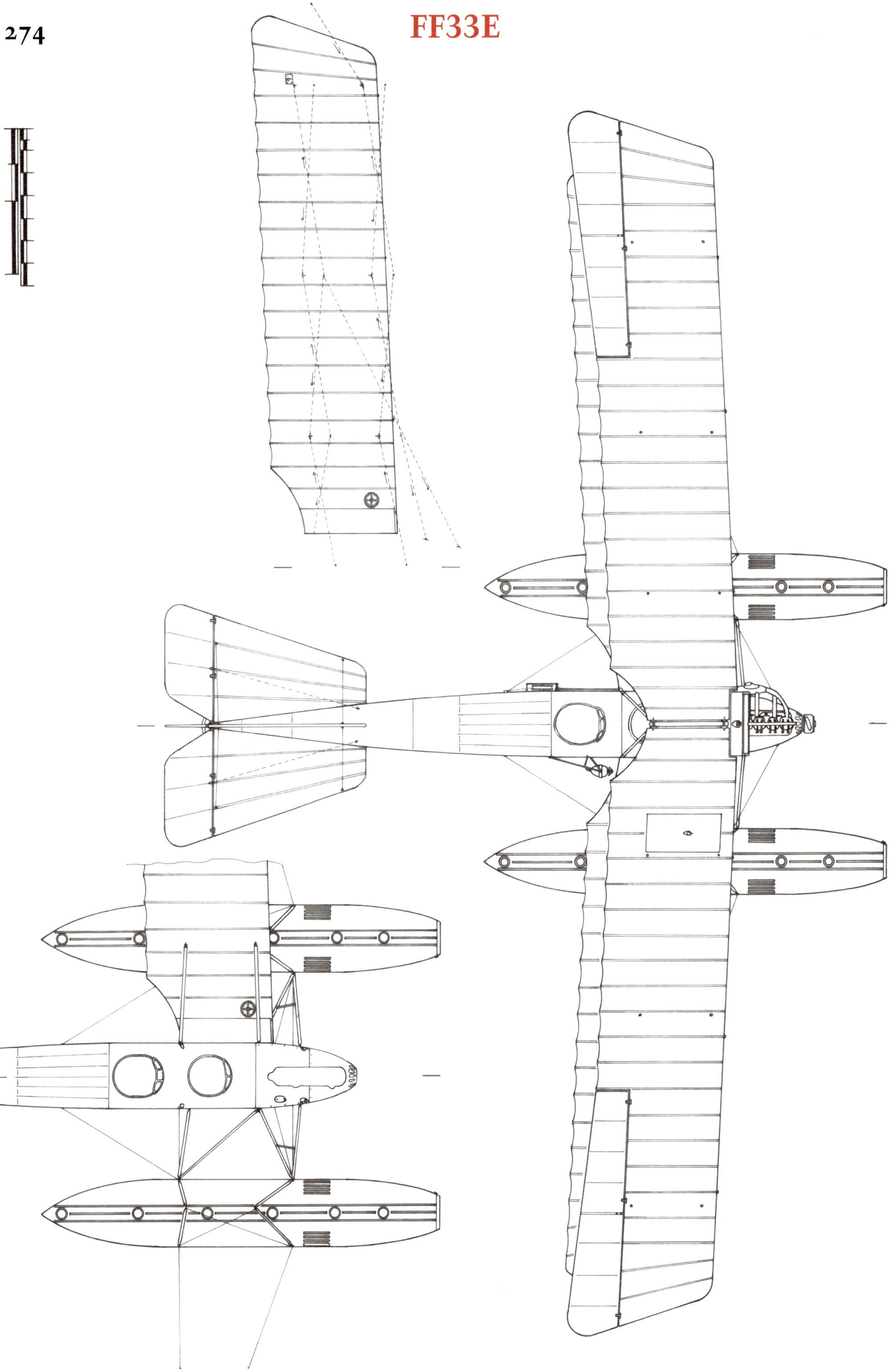

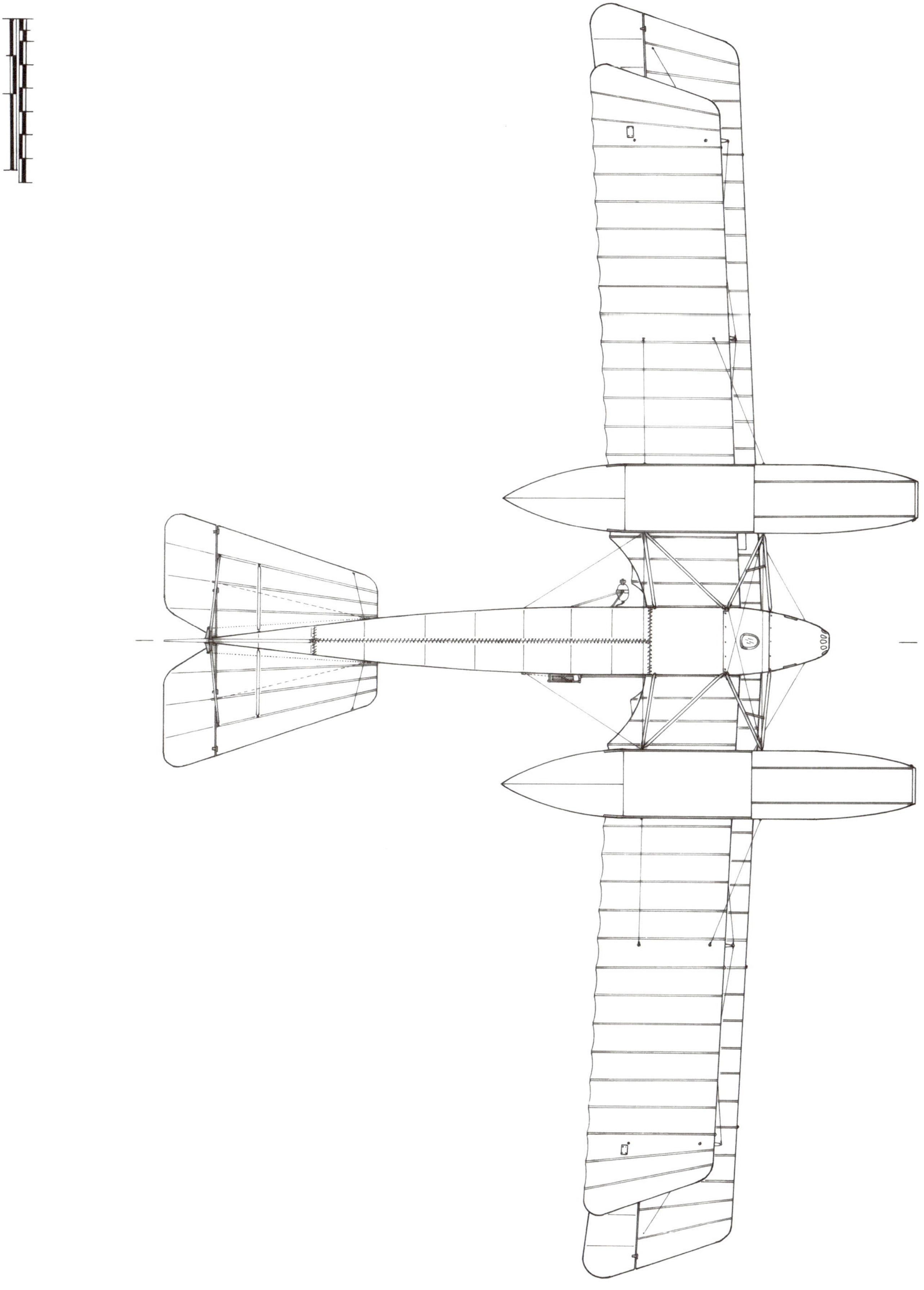

FF33E

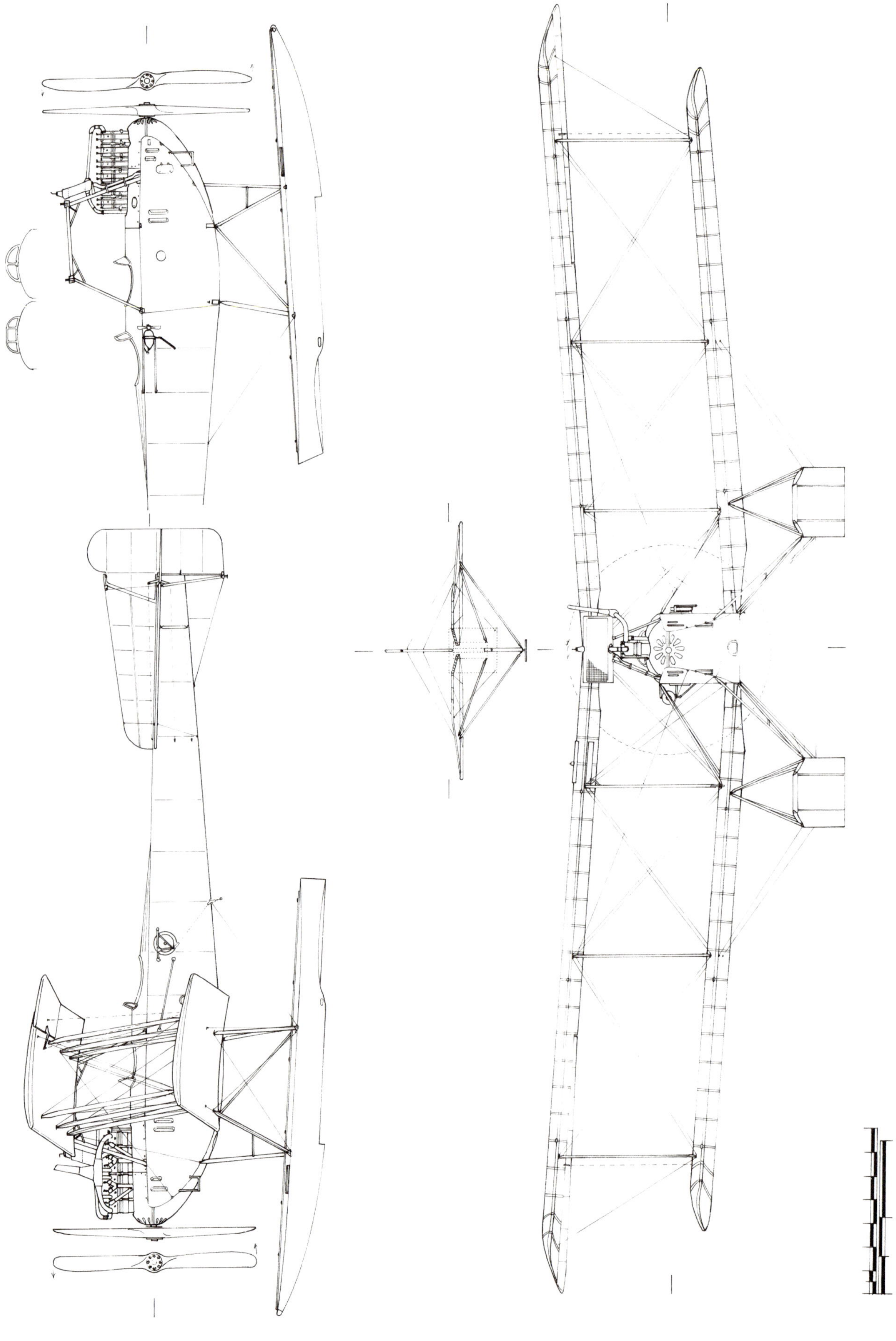

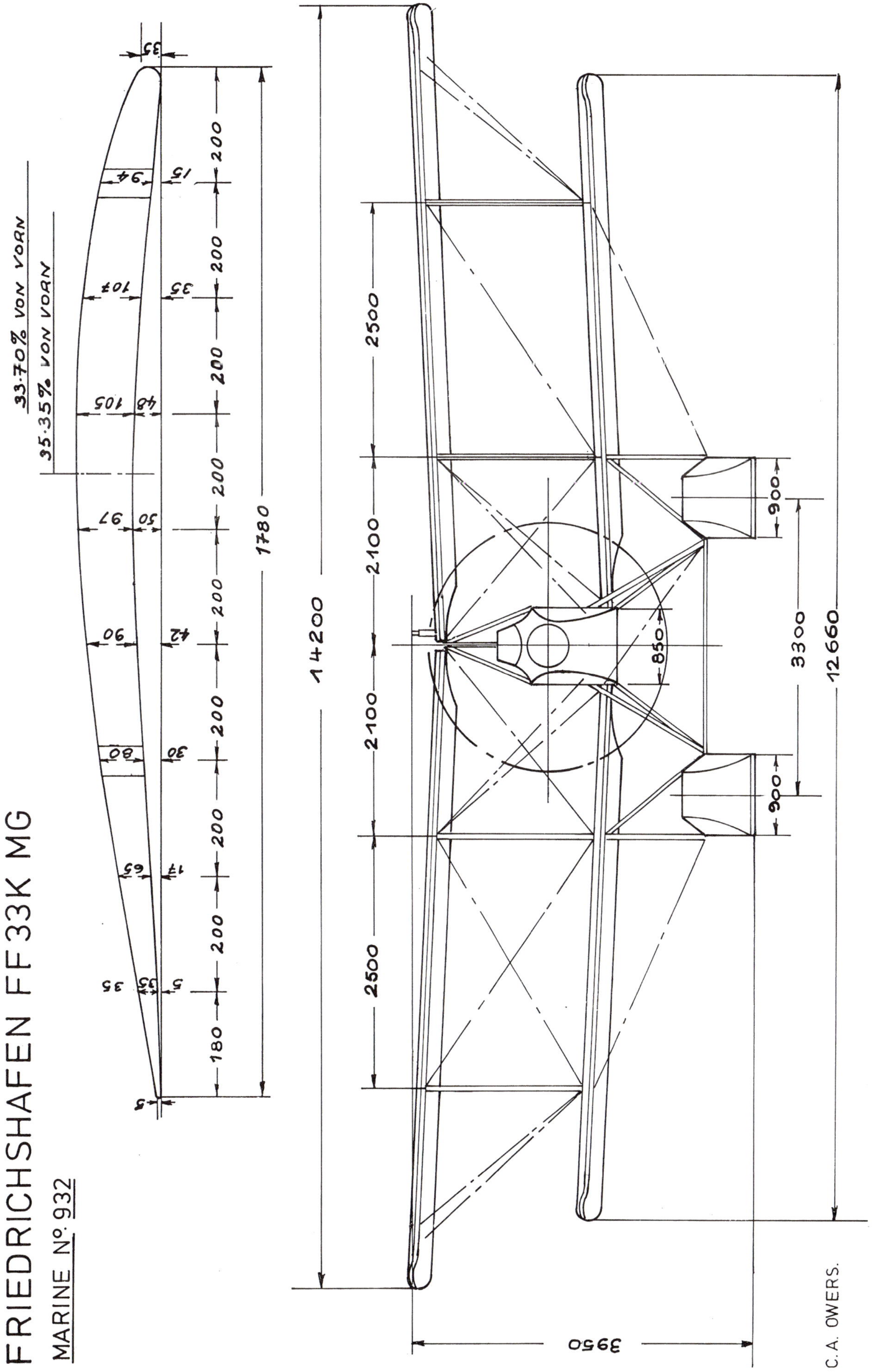
FRIEDRICHSHAFEN FF 33K MG
MARINE Nº 932
33·70% VON VORN
35·35% VON VORN
1780
14200
12650
3950
3300
2500
2100
2100
2500
900
900
850
C.A. OWERS.

FF33K

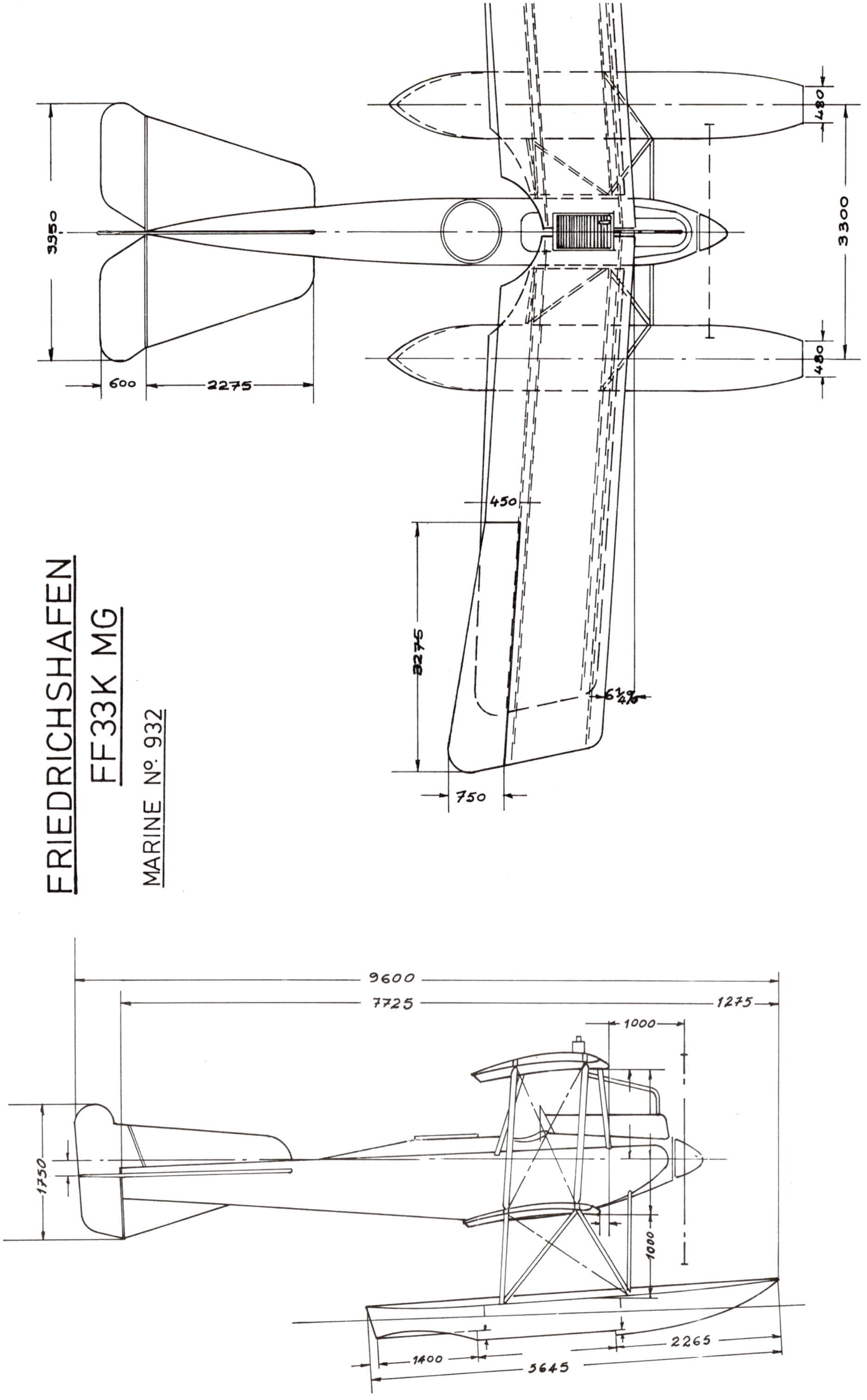
FRIEDRICHSHAFEN
FF33K MG
MARINE No. 932
3350
600
2275
480
3300
480
450
3275
750
9600
7725
1275
1000
1750
1000
2265
1400
3645

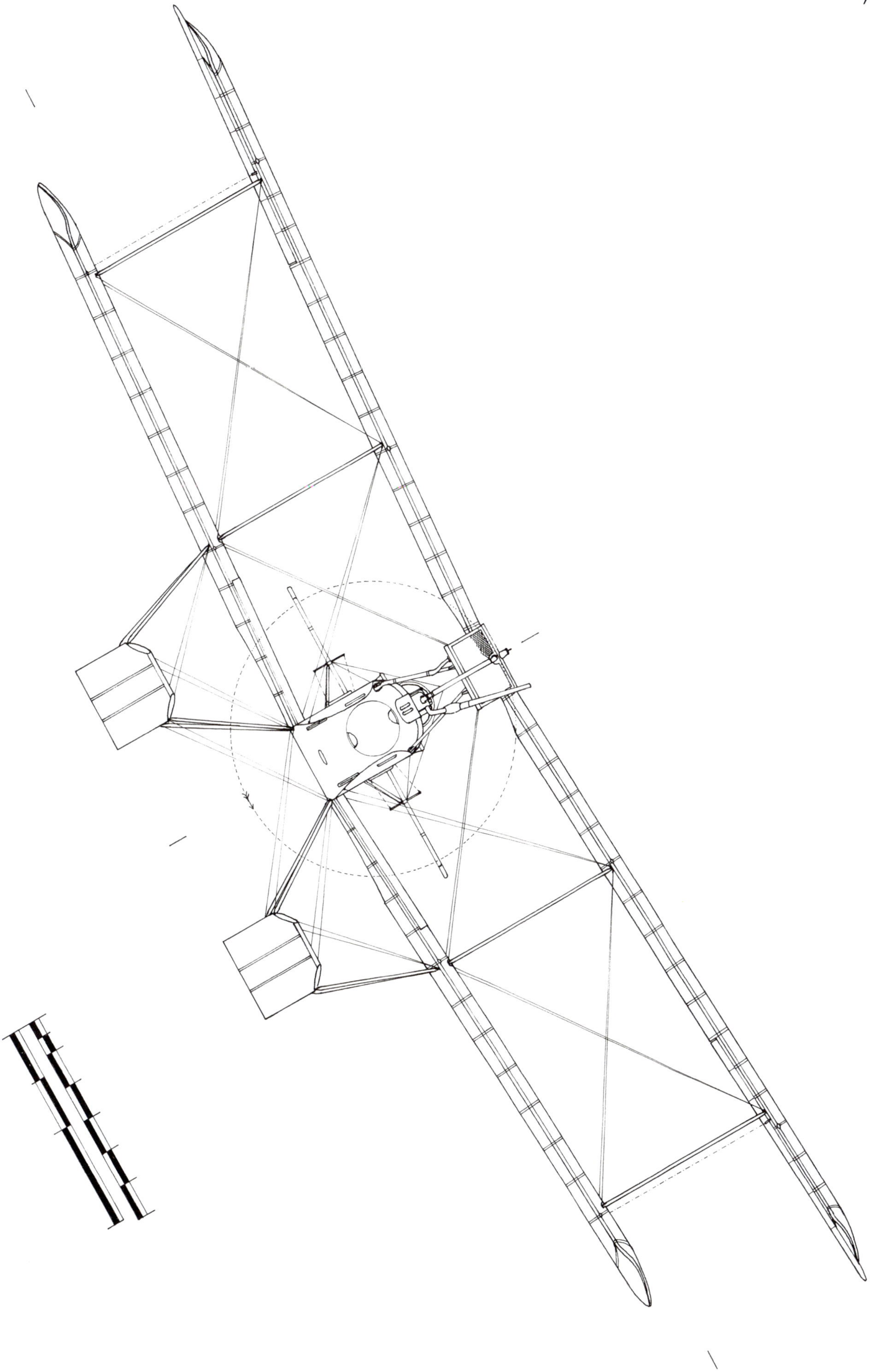

FF33L

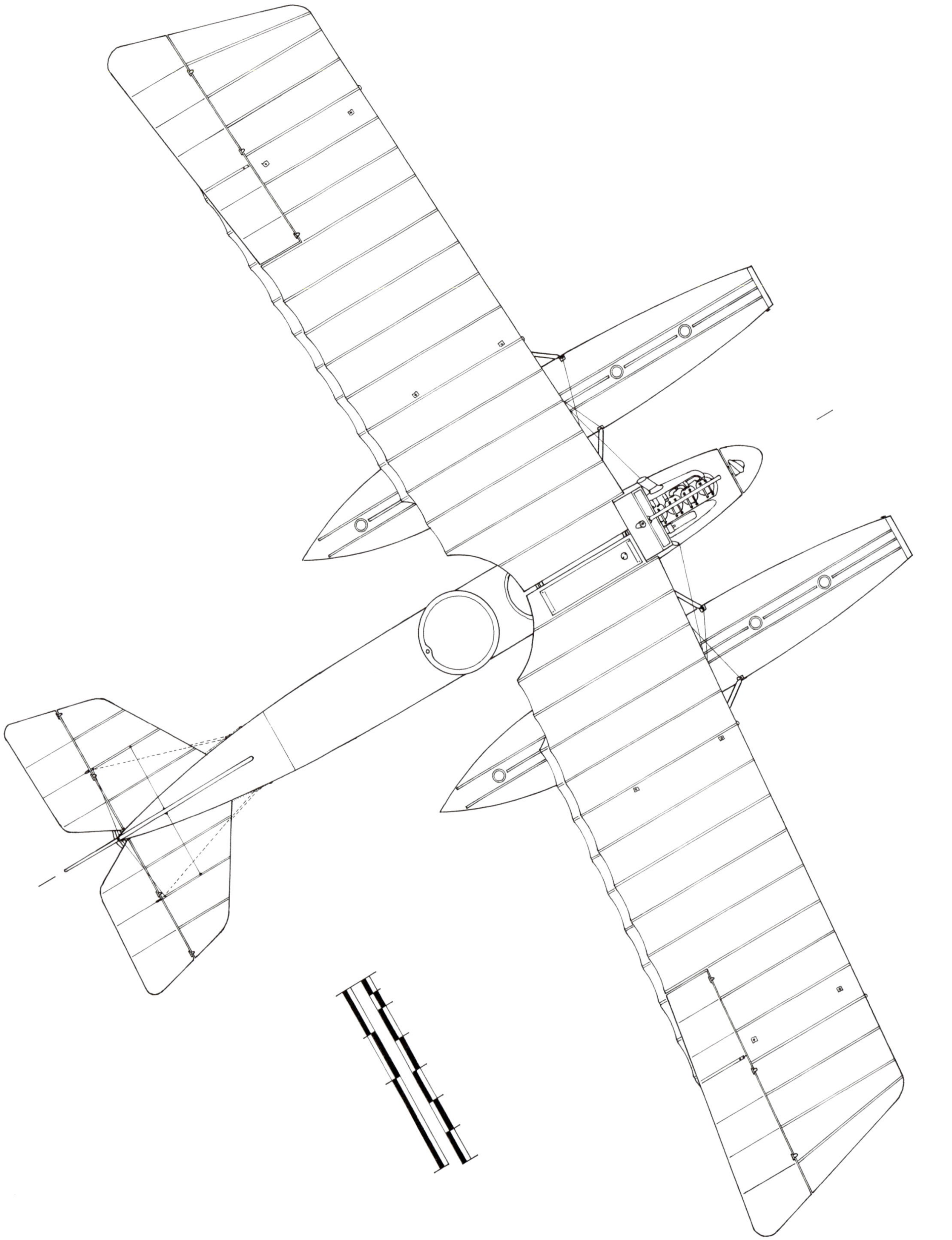

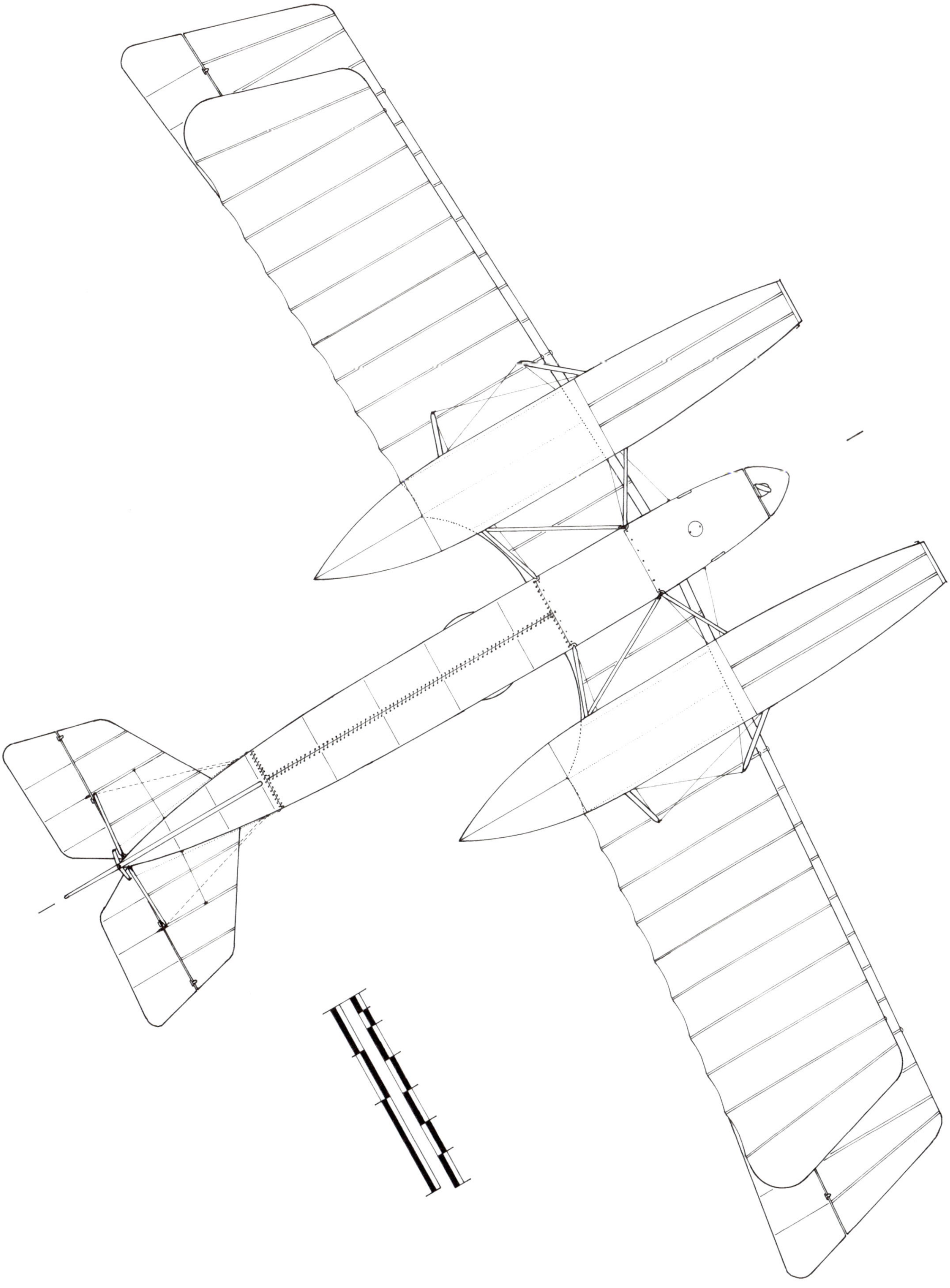

FF33L

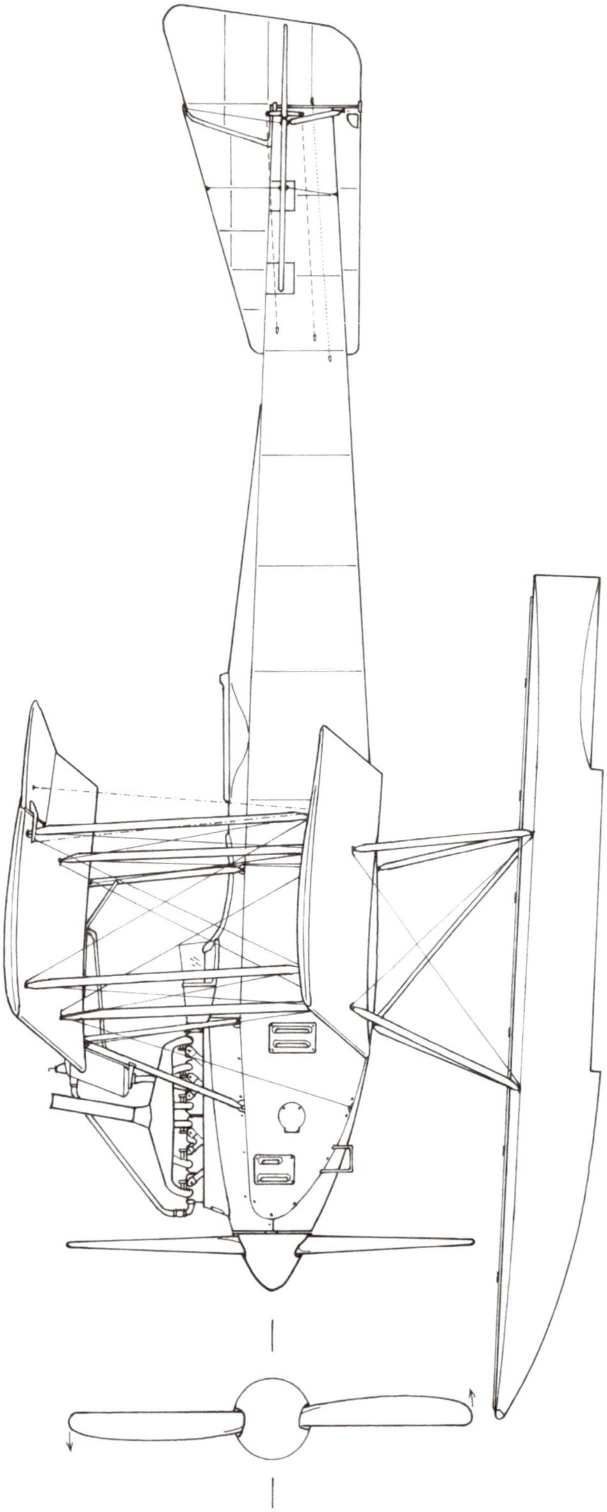

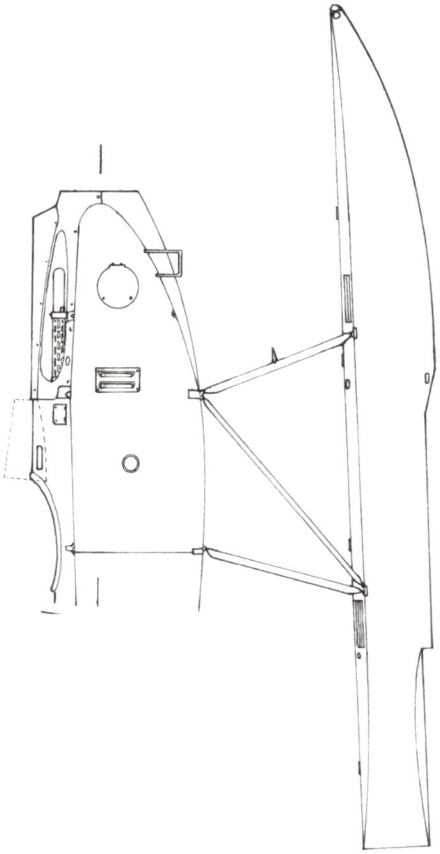

FF43

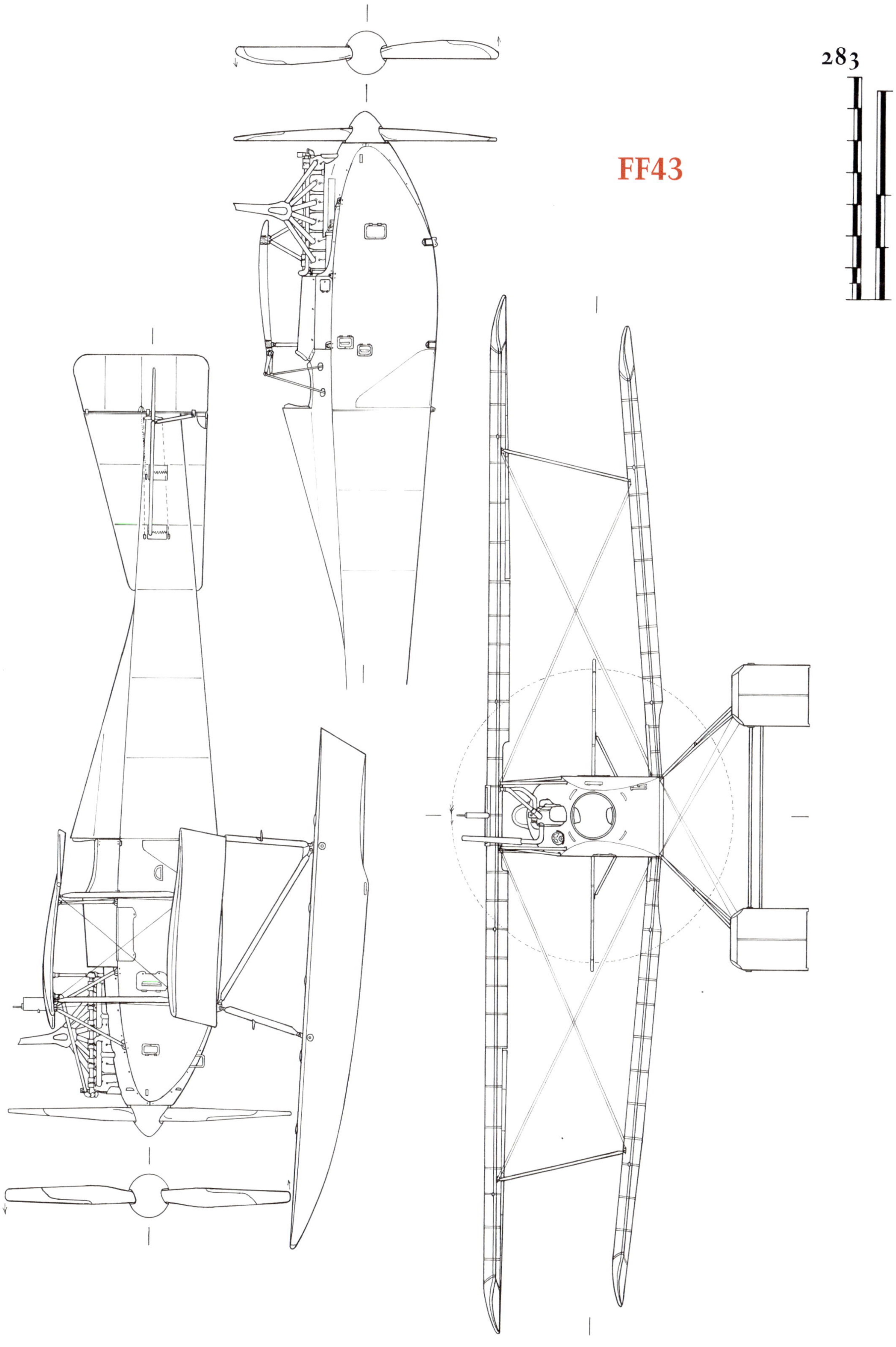

FF43

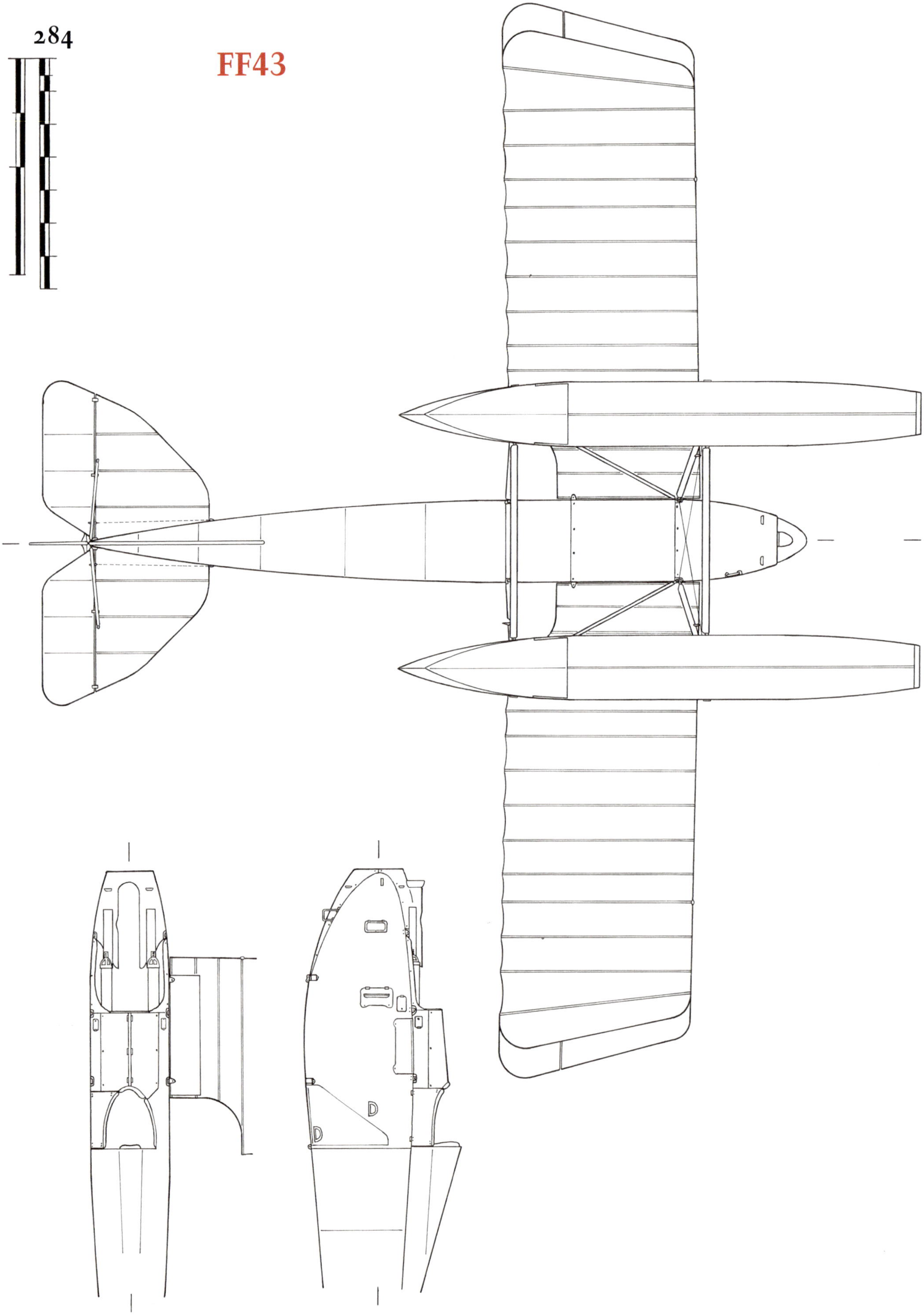

FF43

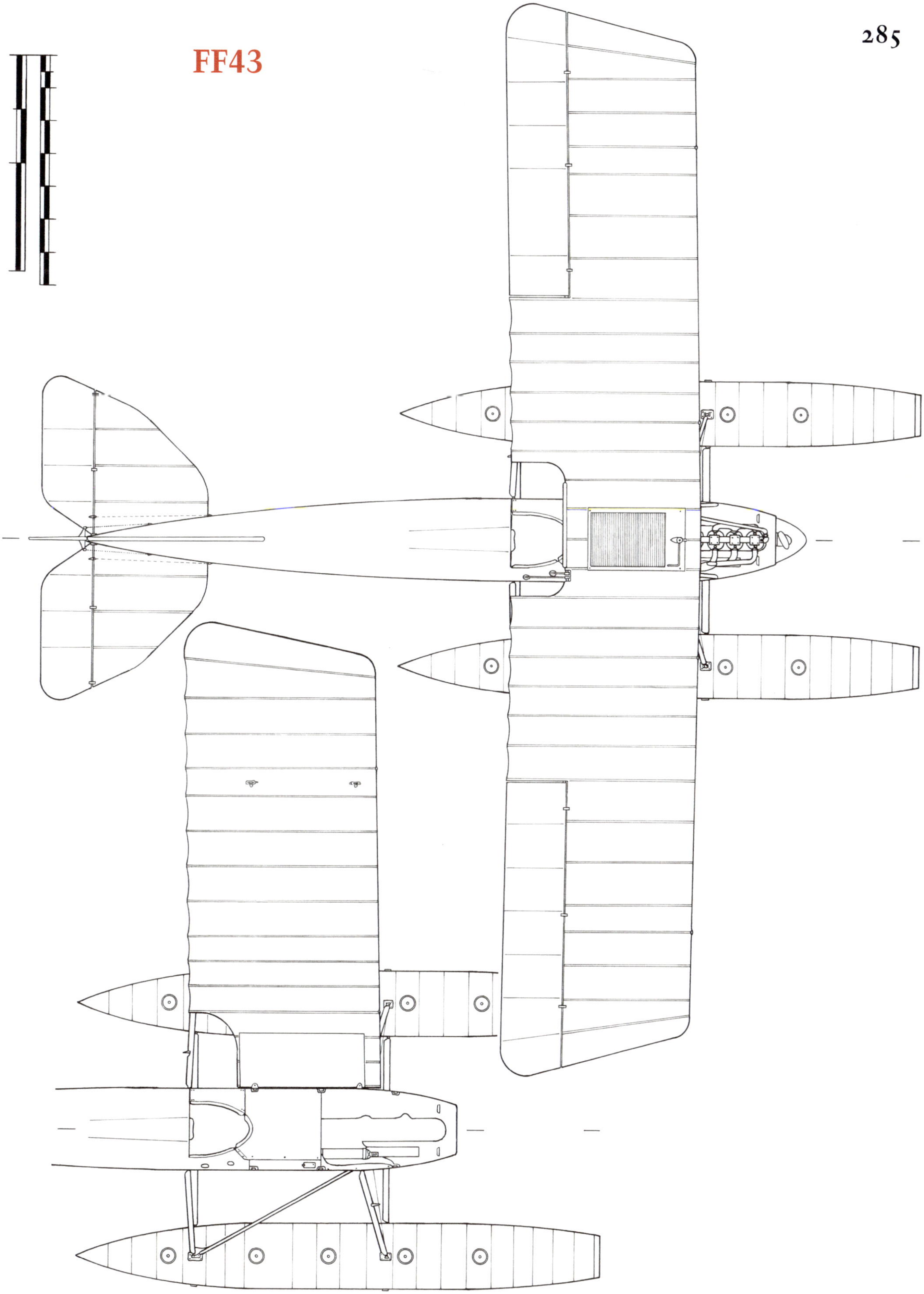

FF39

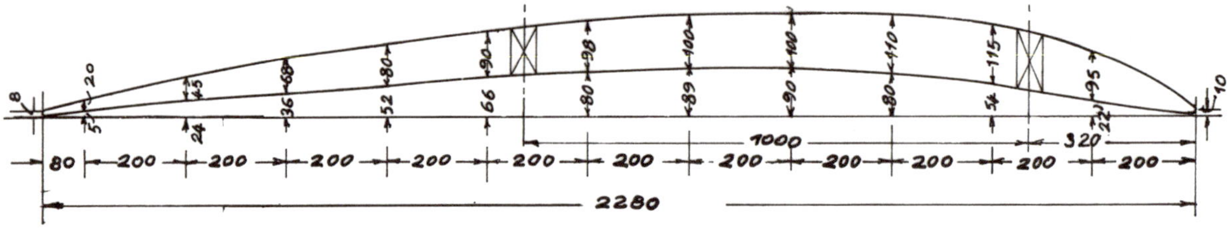
2280
1000
320

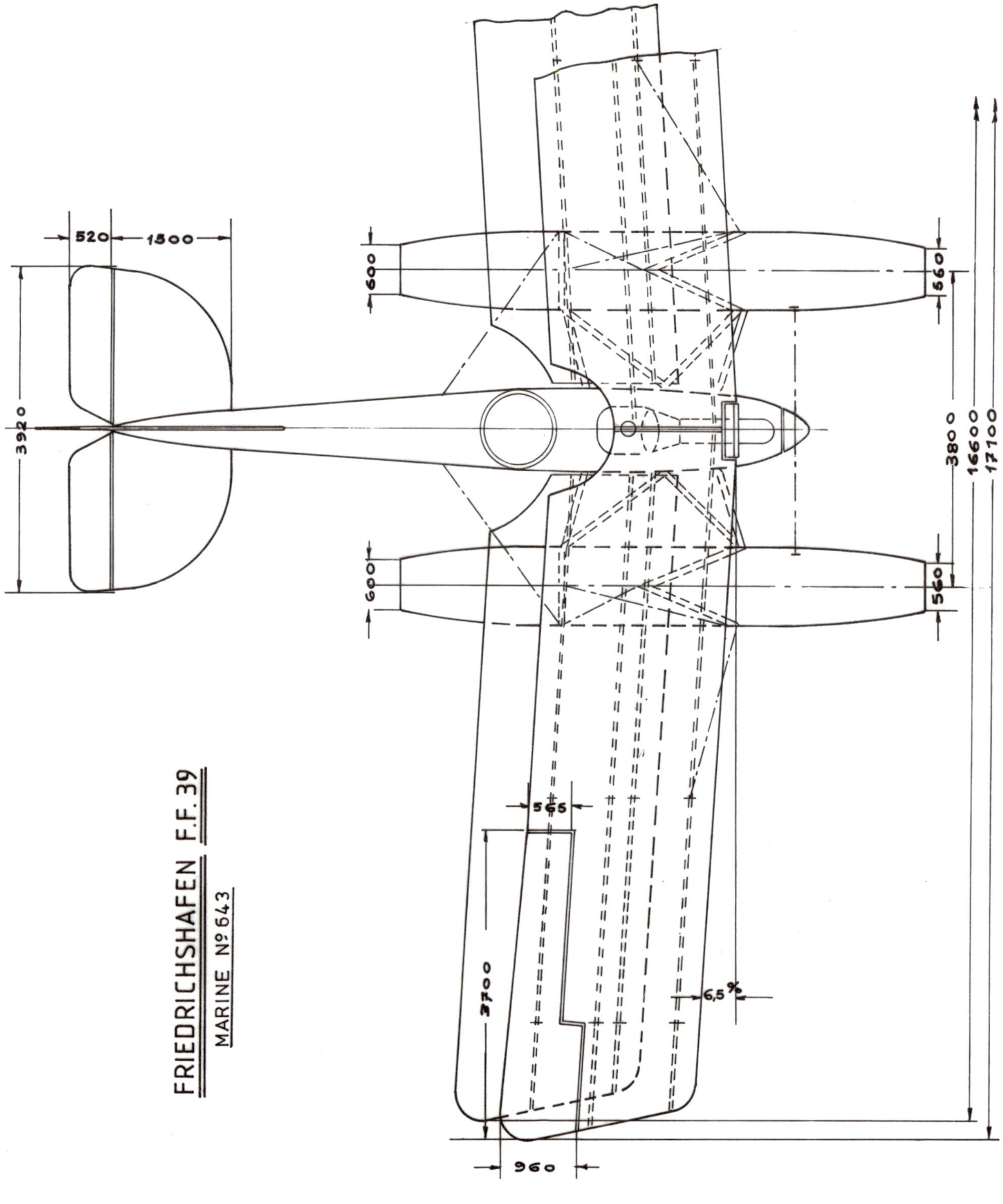
FRIEDRICHSHAFEN F.F. 39
MARINE No 643
520
1500
3920
600
560
3800
16600
17100
565
3700
960
6,5 %

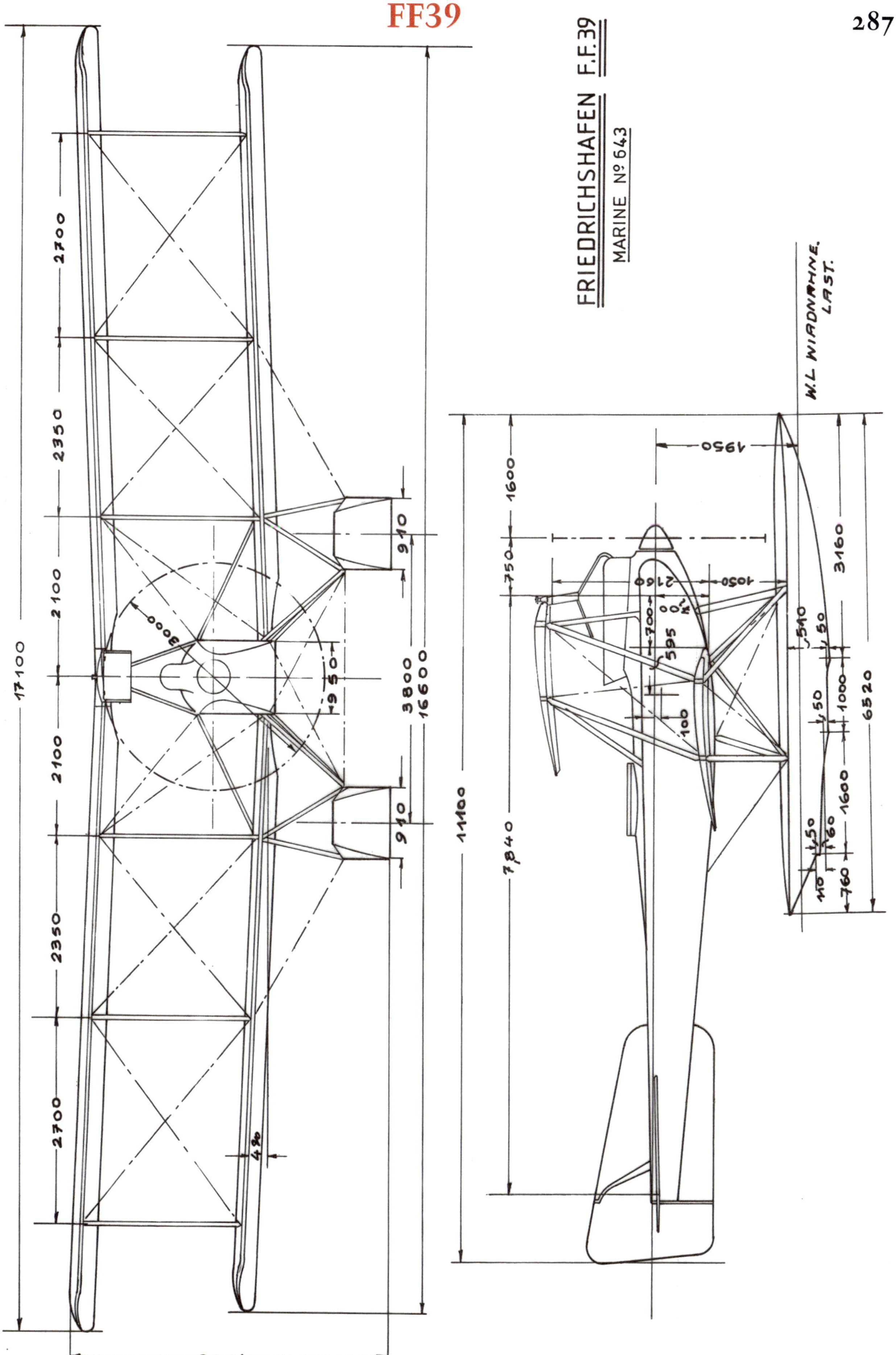
FRIEDRICHSHAFEN F.F.39
MARINE Nº 643

FF41A

MN 678

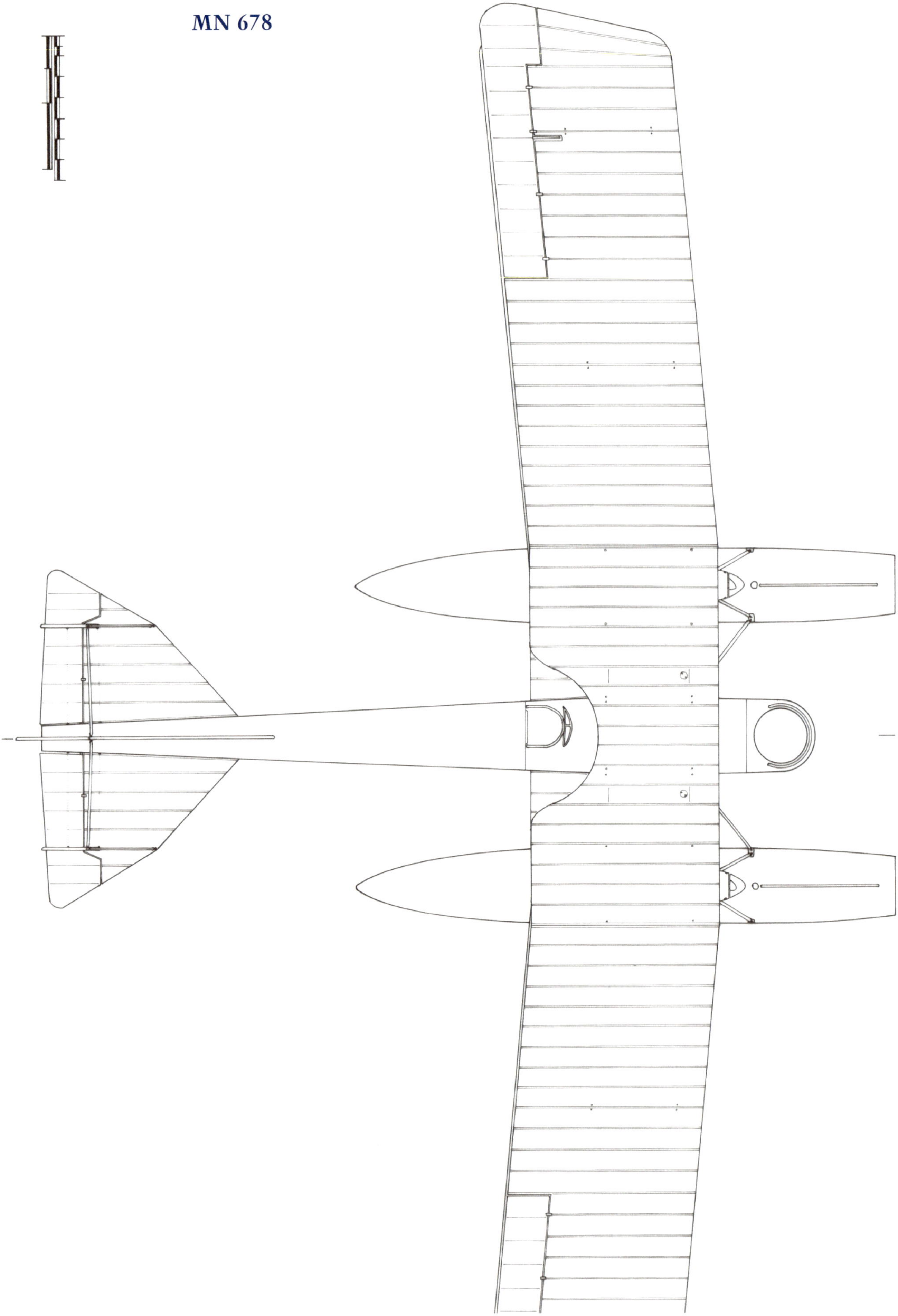

MN 678

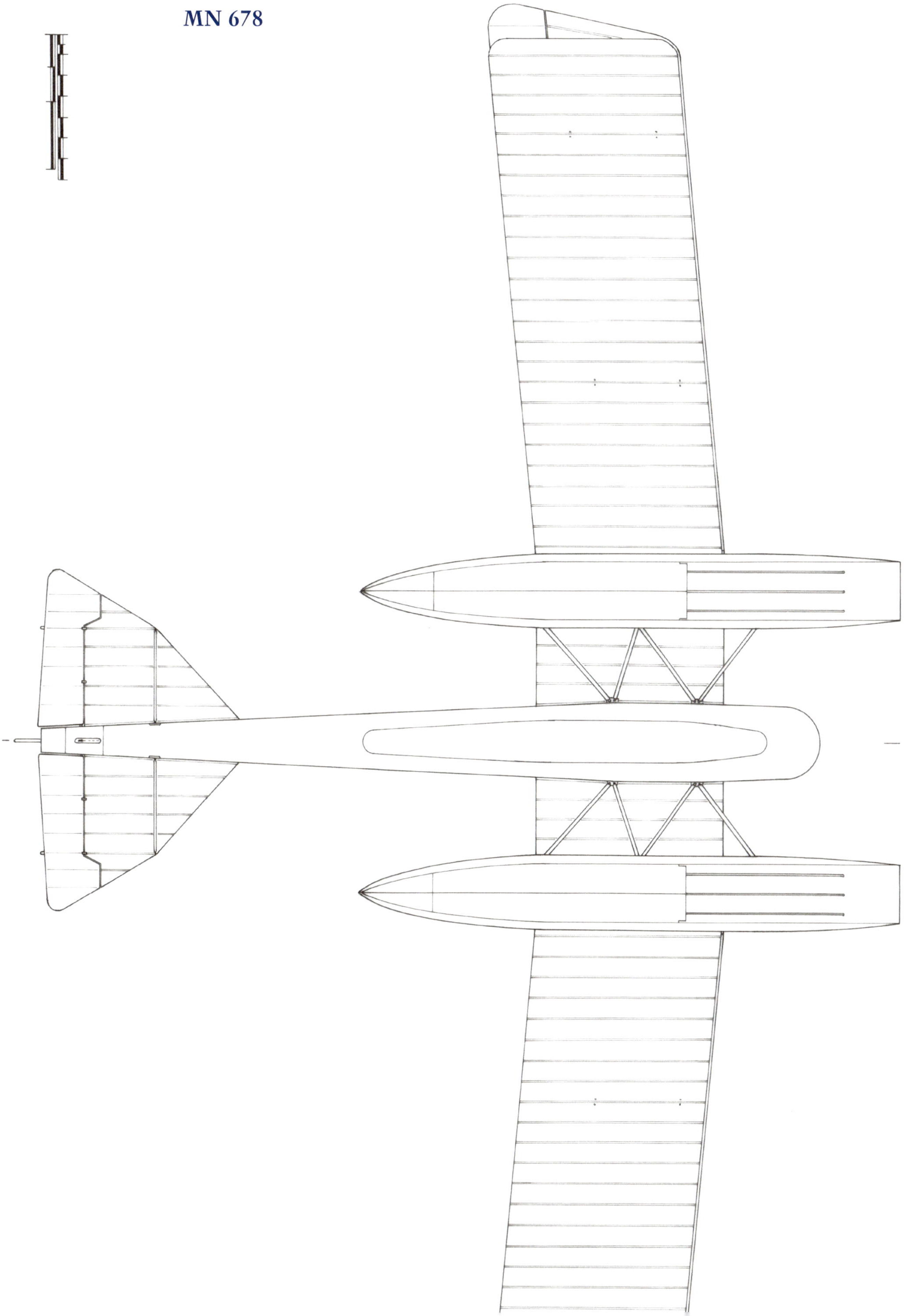

FF41A

MN 678

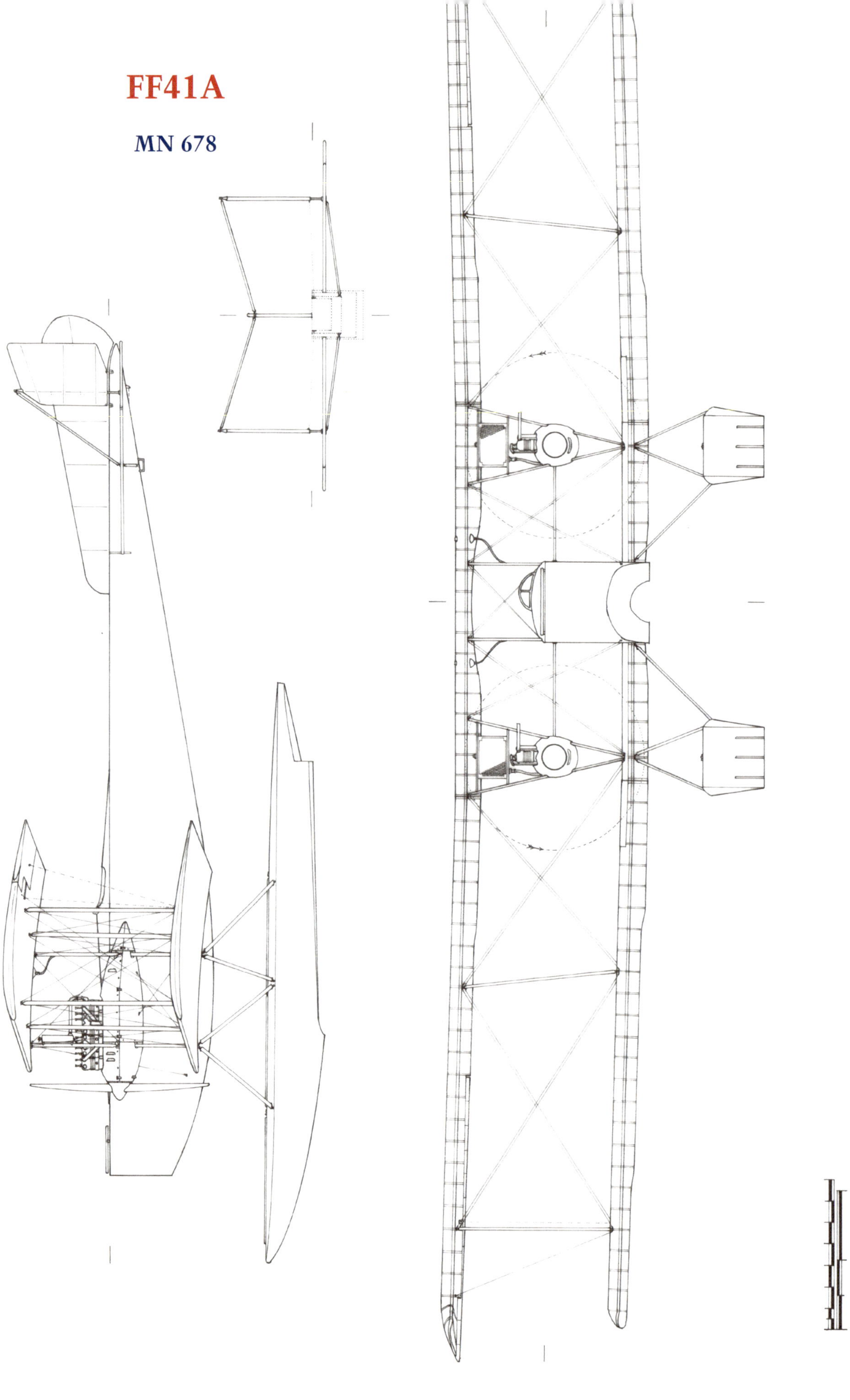

FF41AT

MN 996–1000 & 1208–1210

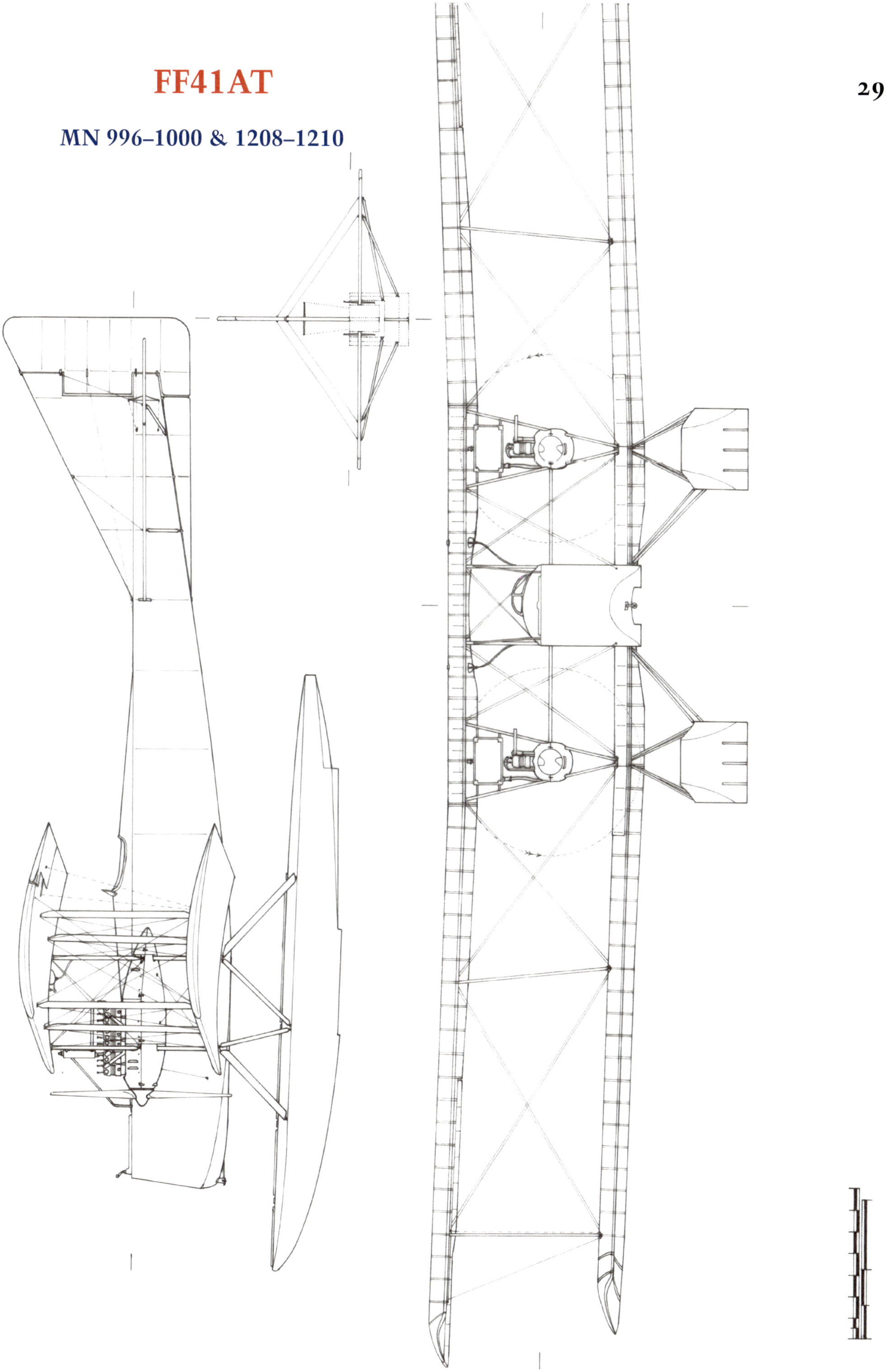

FF41AT

MN 996–1000 & 1208–1210

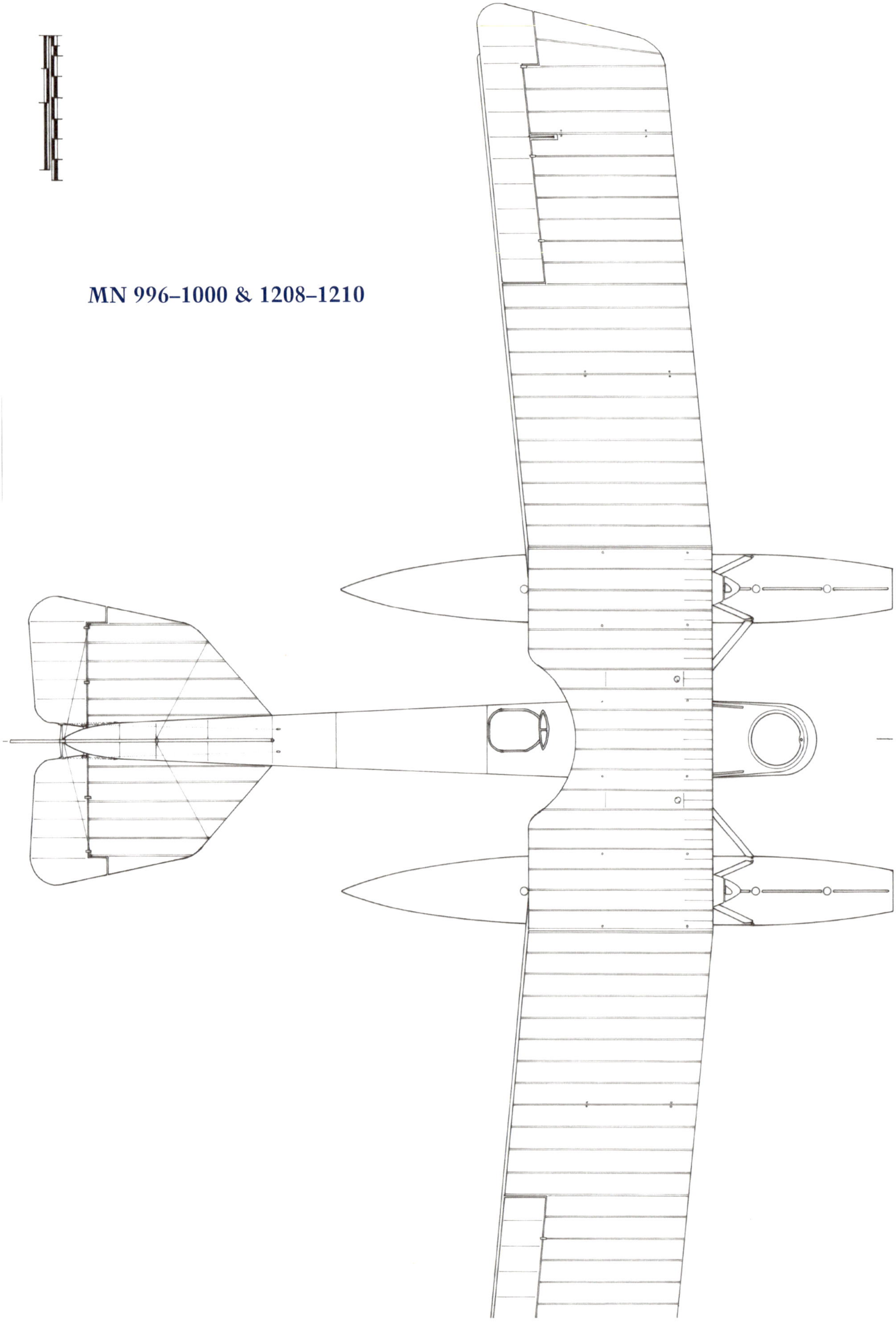

MN 996–1000 & 1208–1210

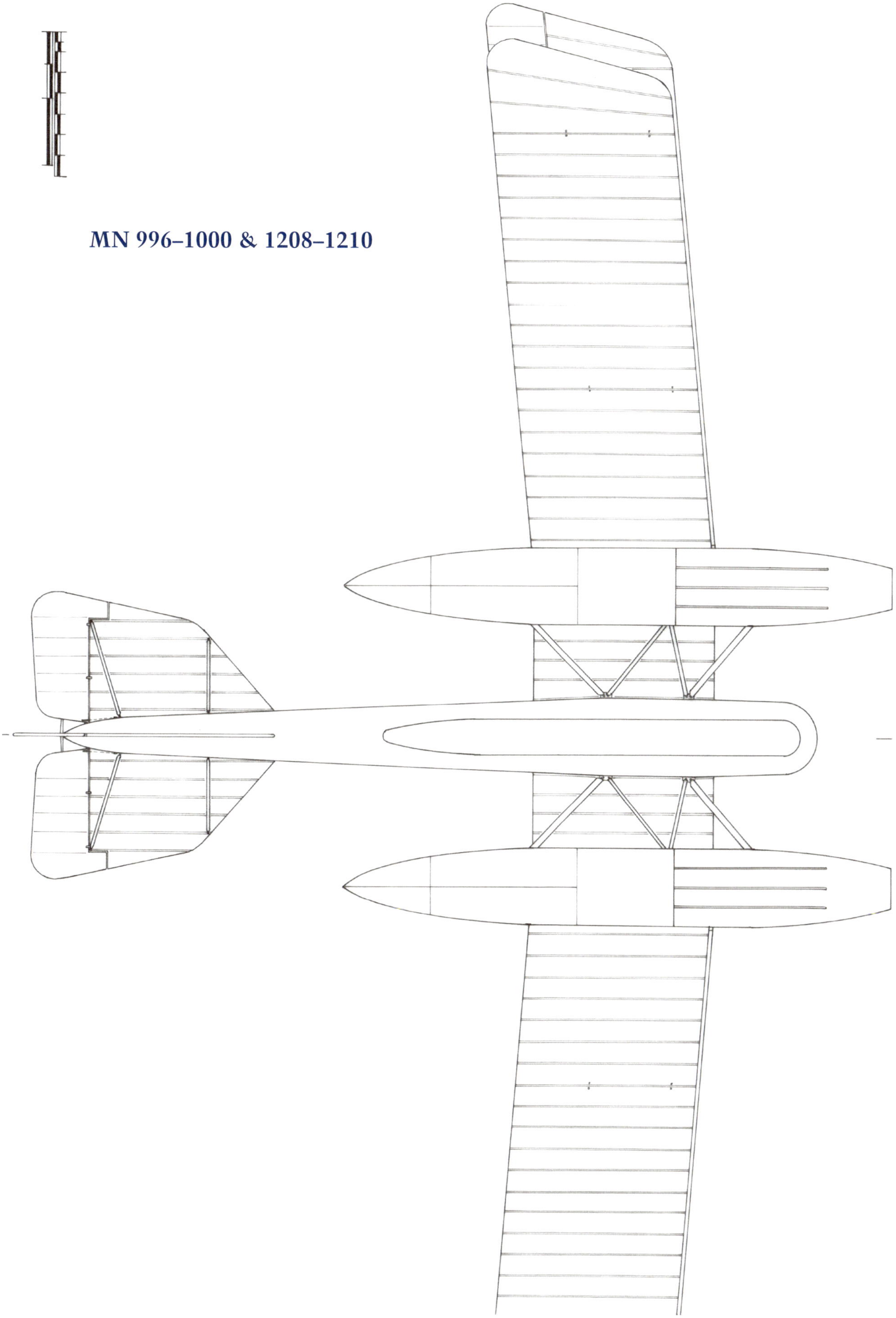

FF49C

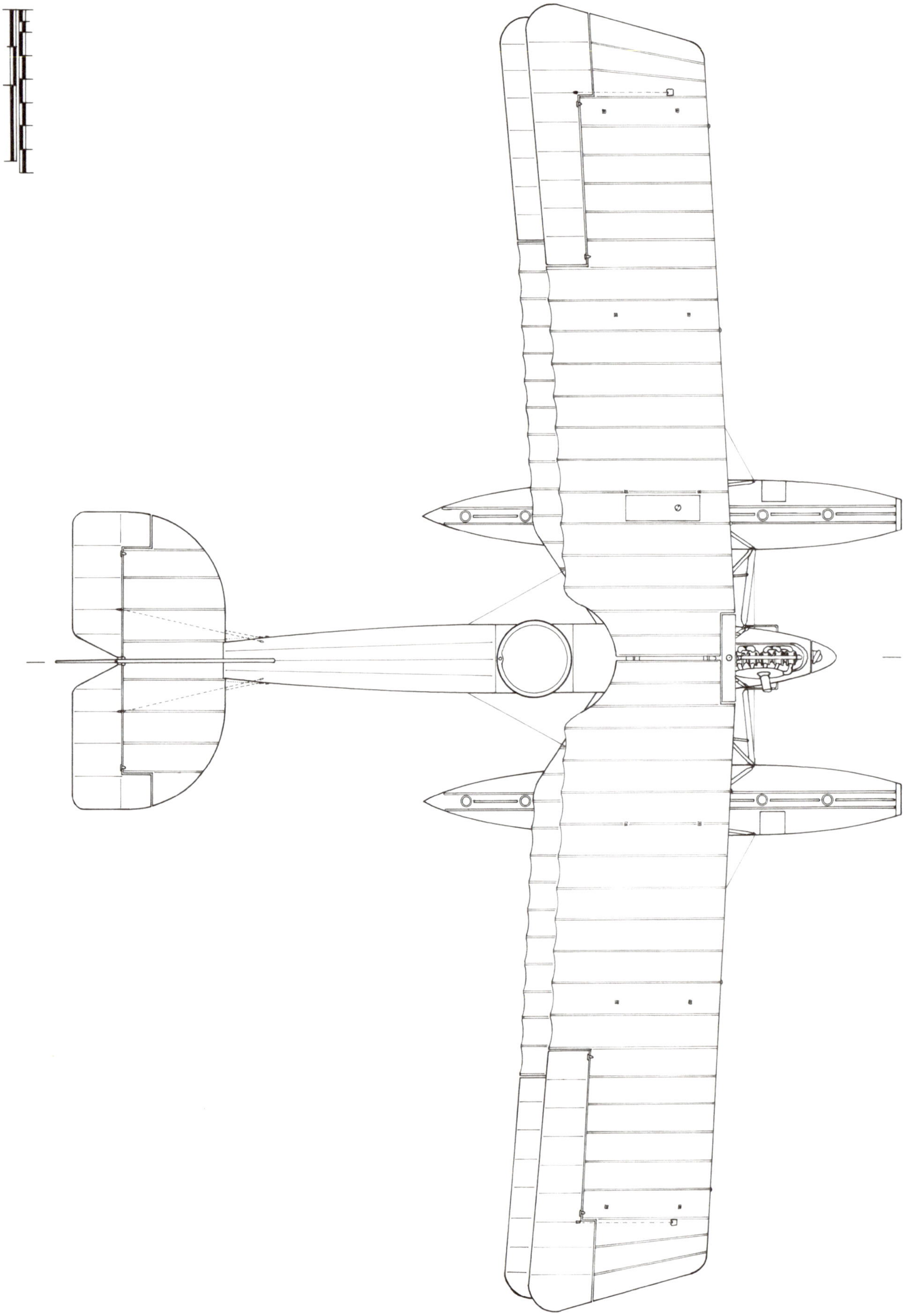

FF49C

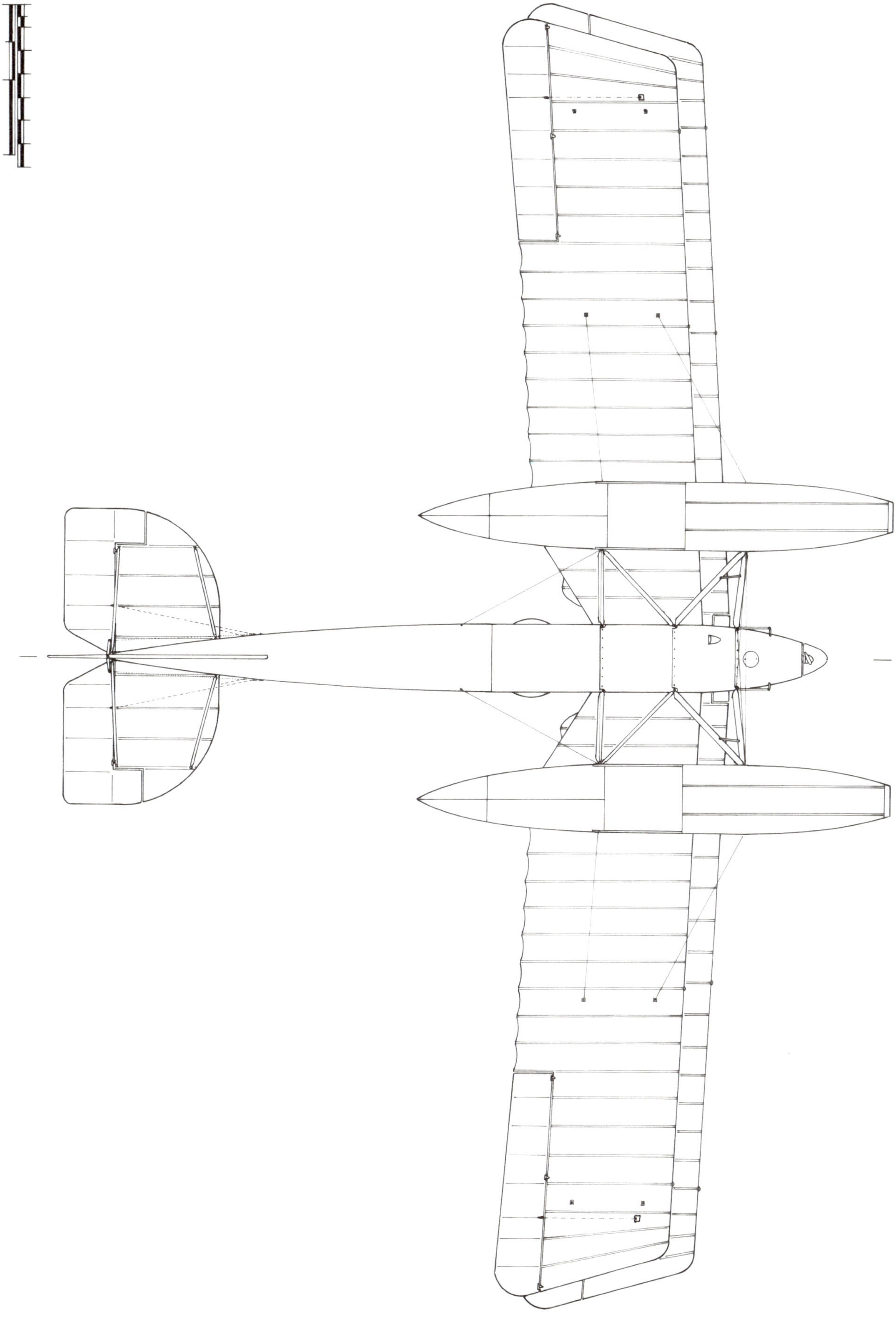

FF49C

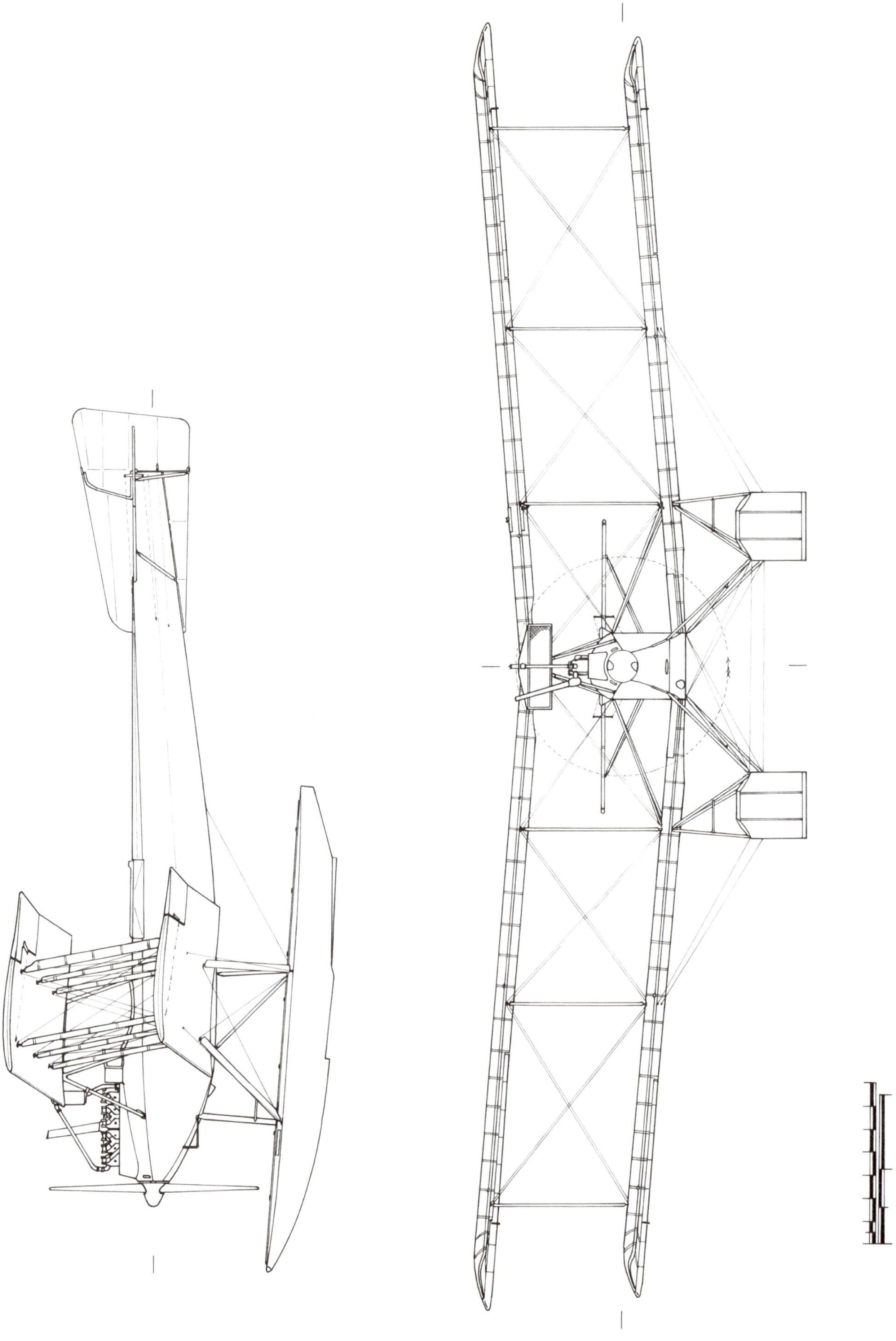

G.II

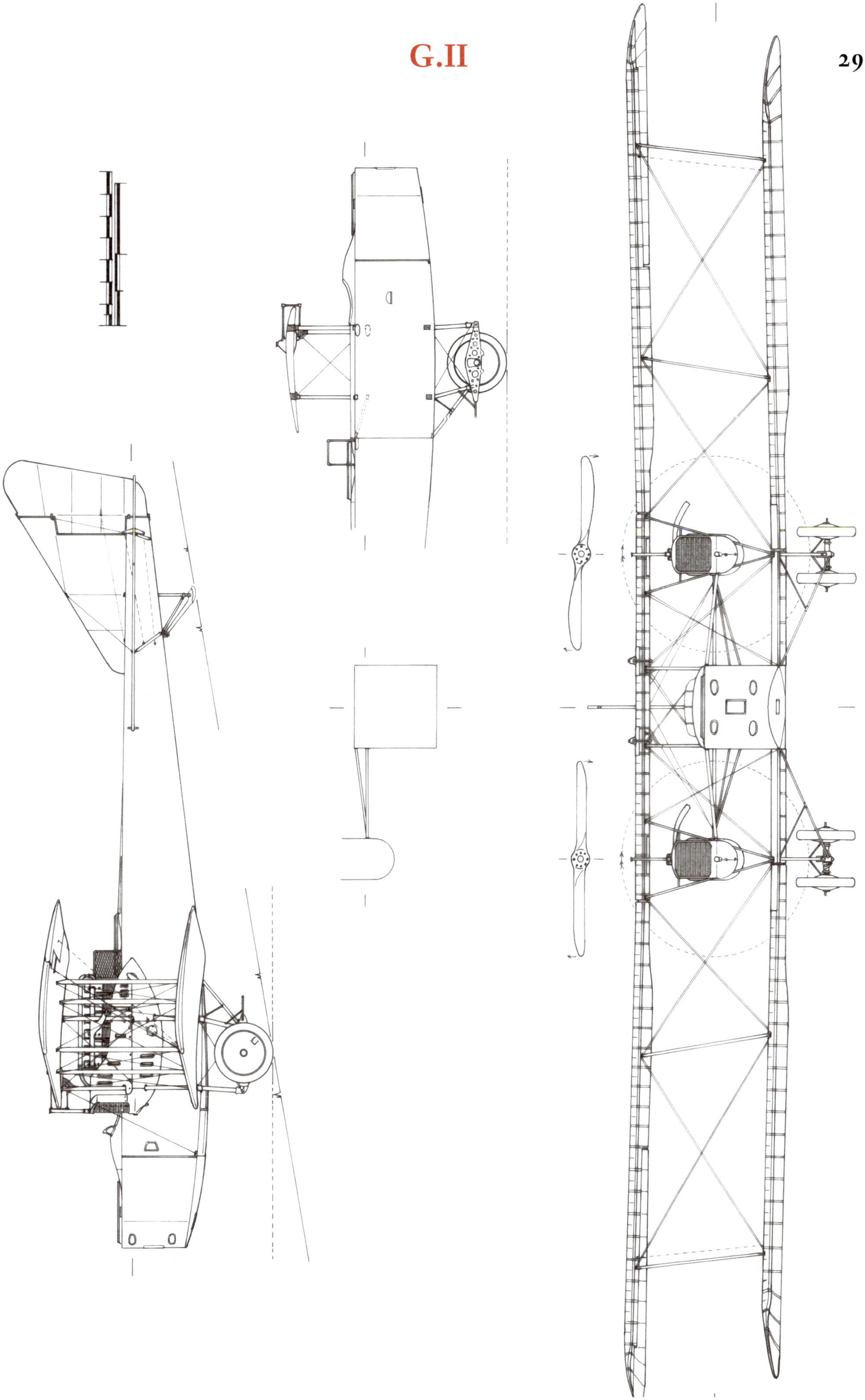

G.II

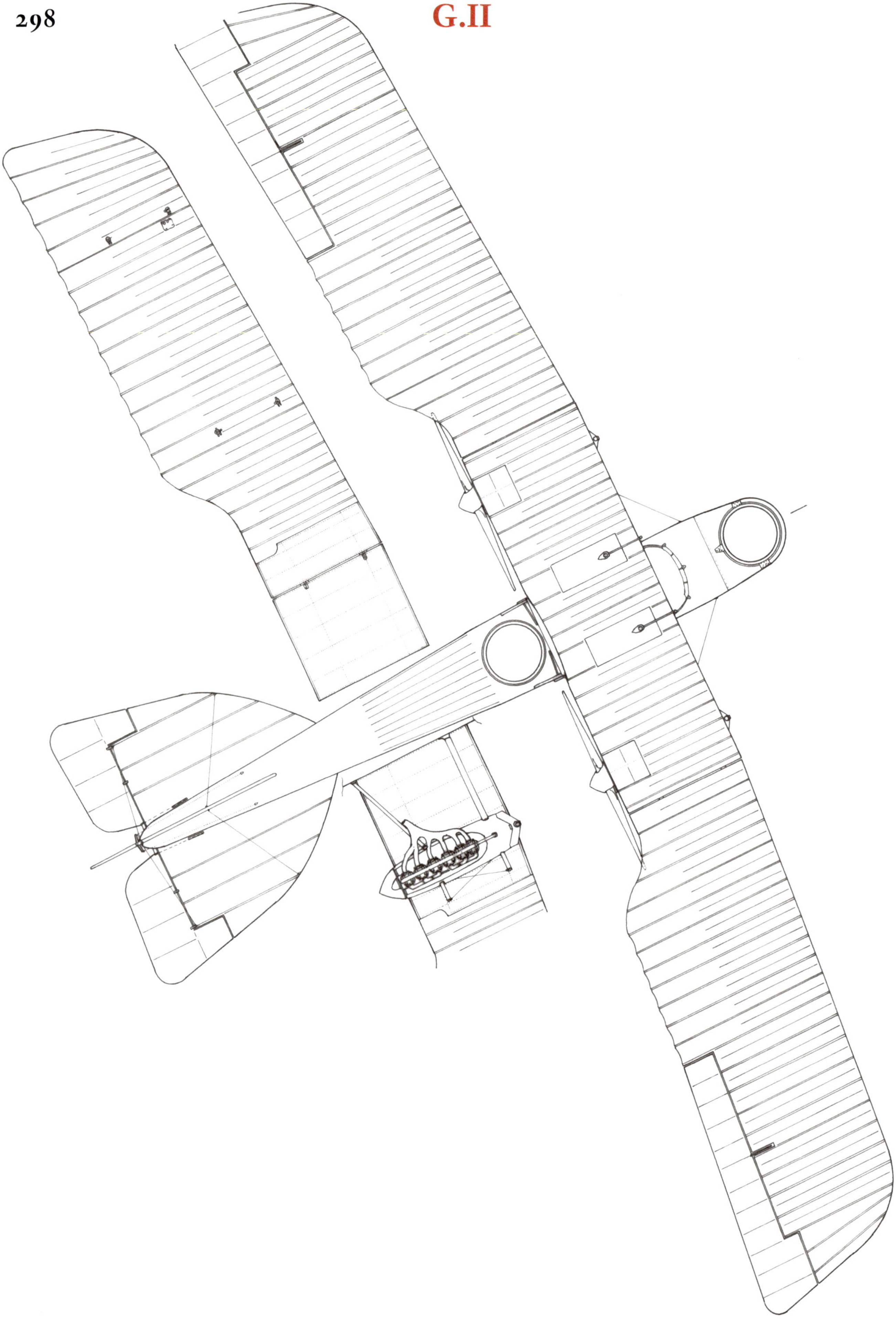

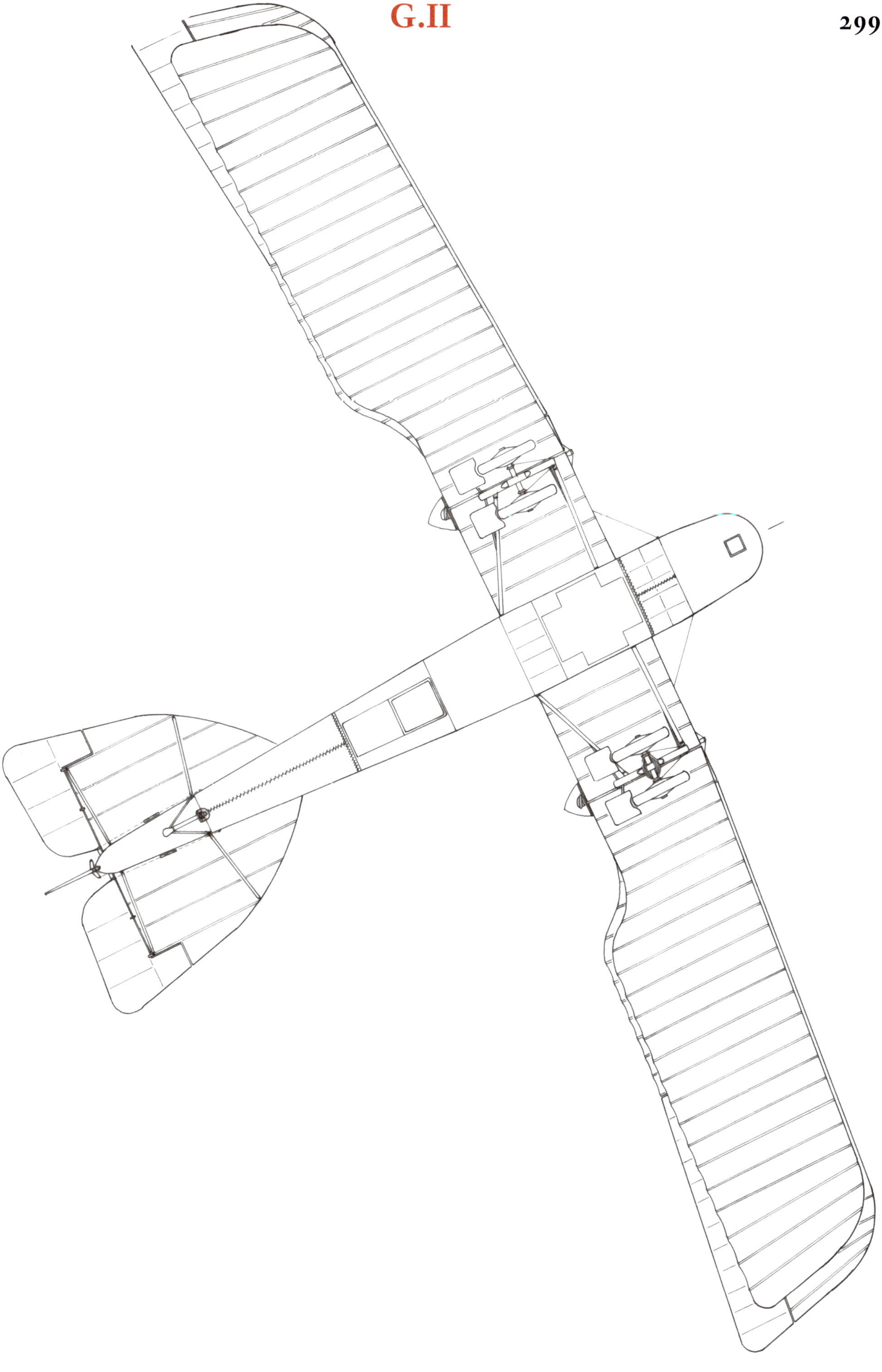

D.I

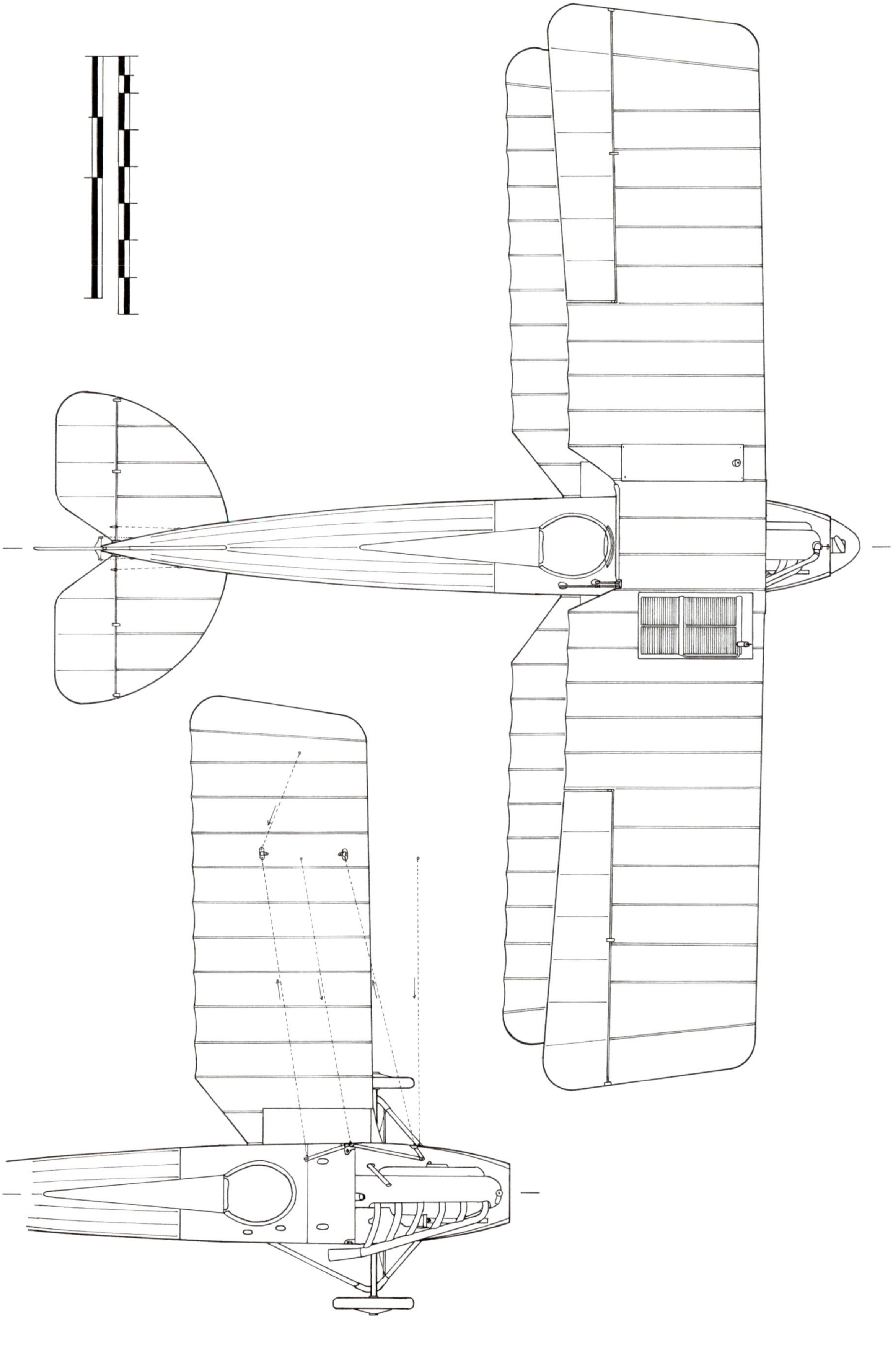

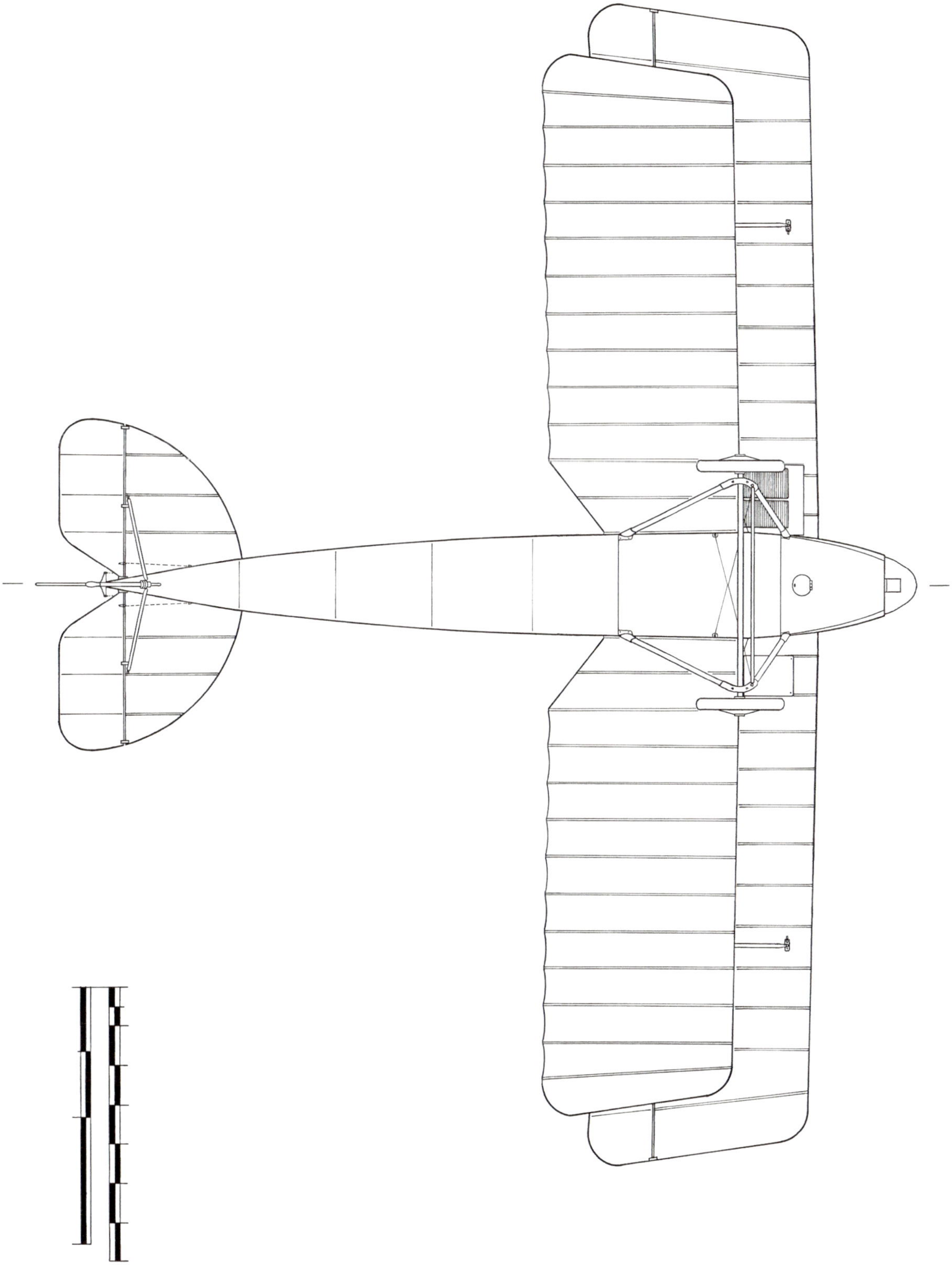

D.I

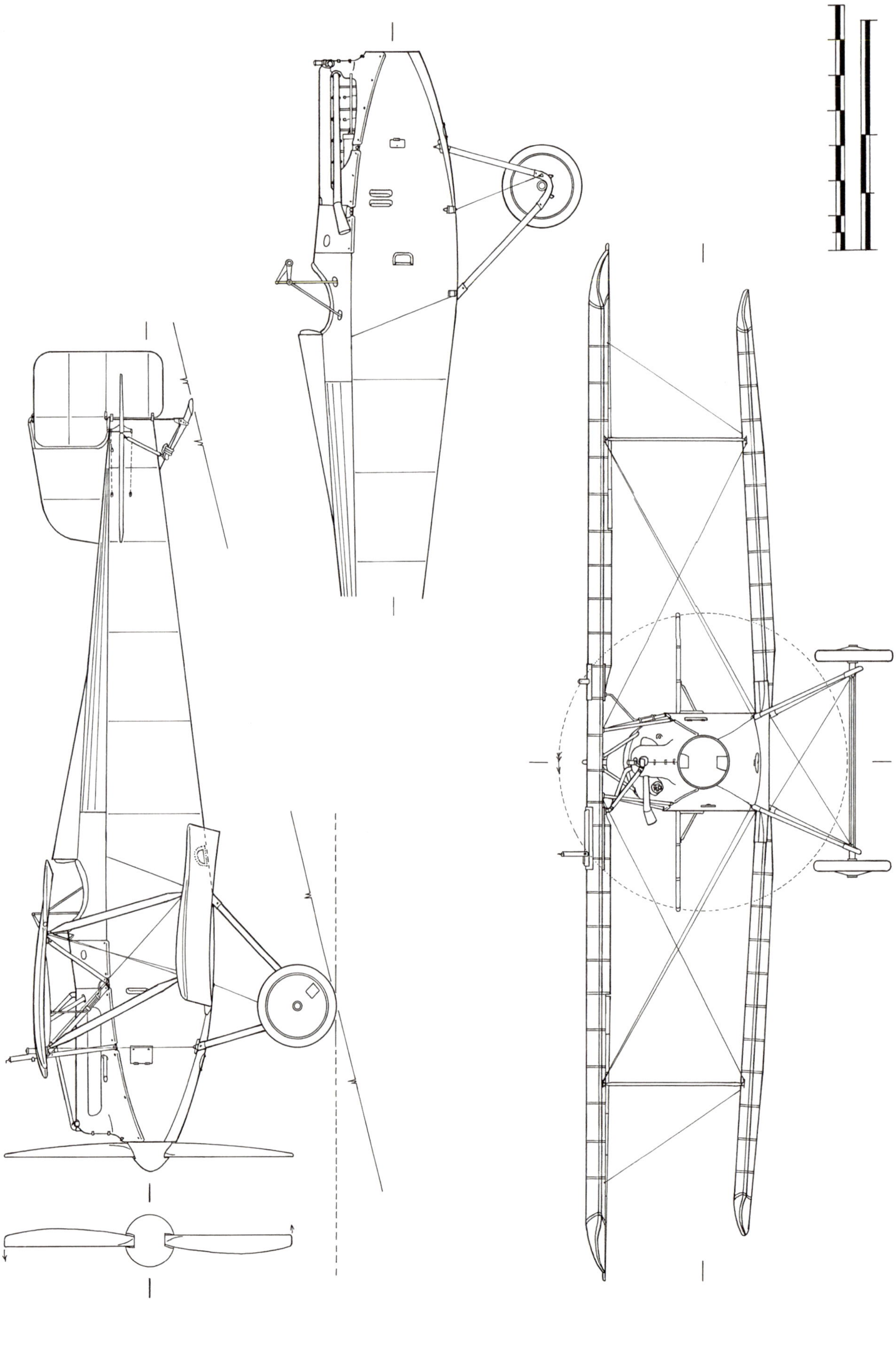

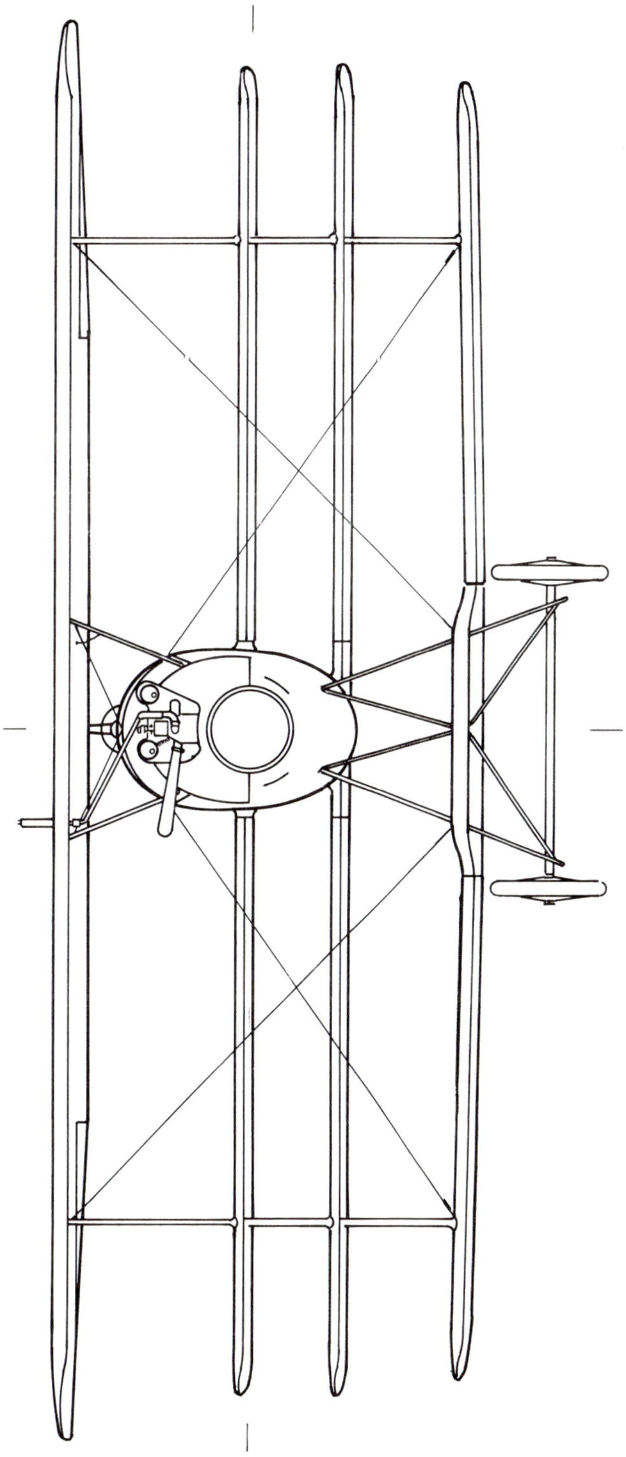

FF54

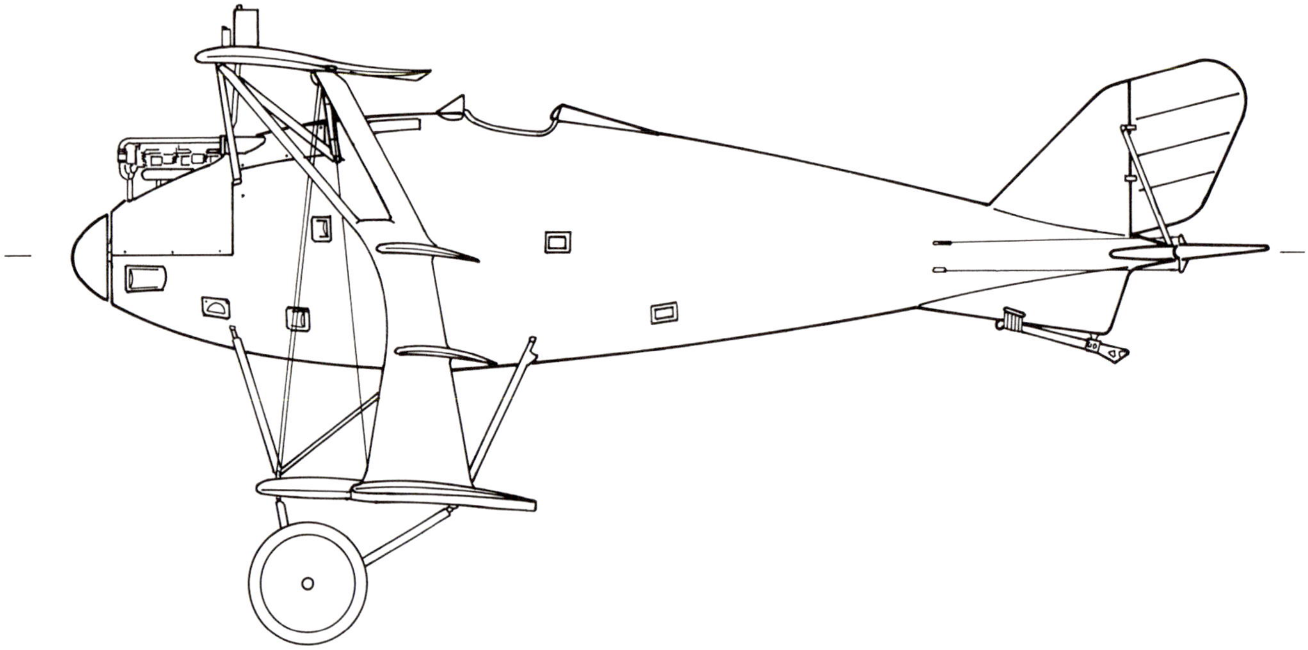

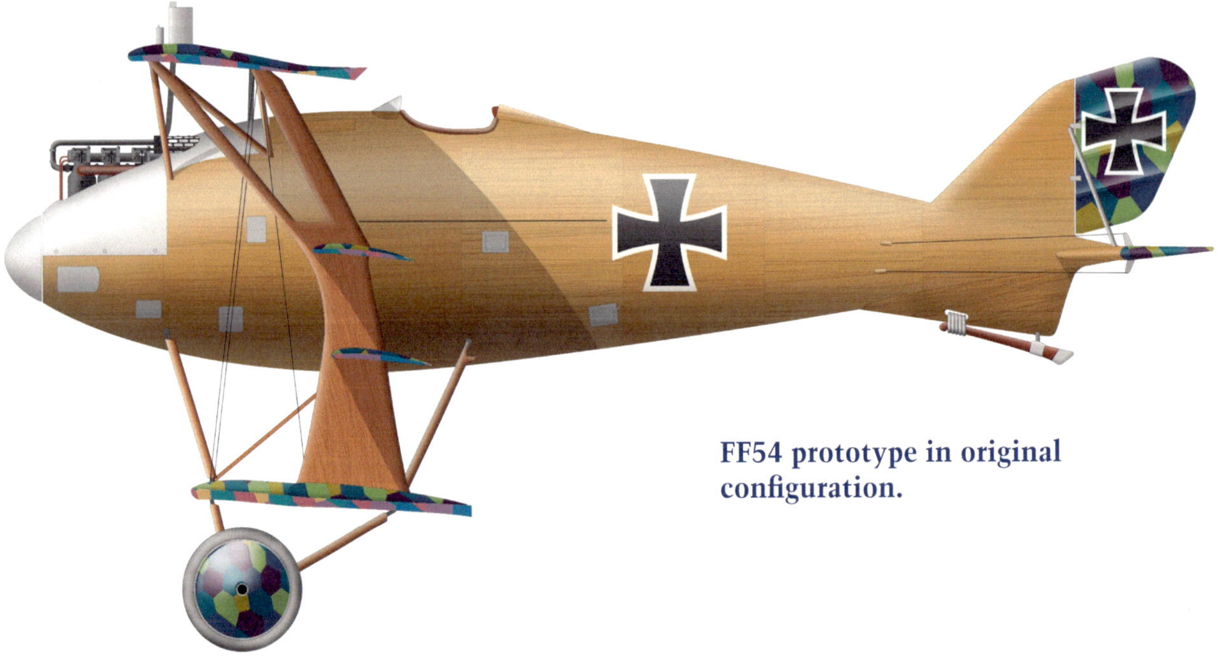

FF54 prototype in original configuration.

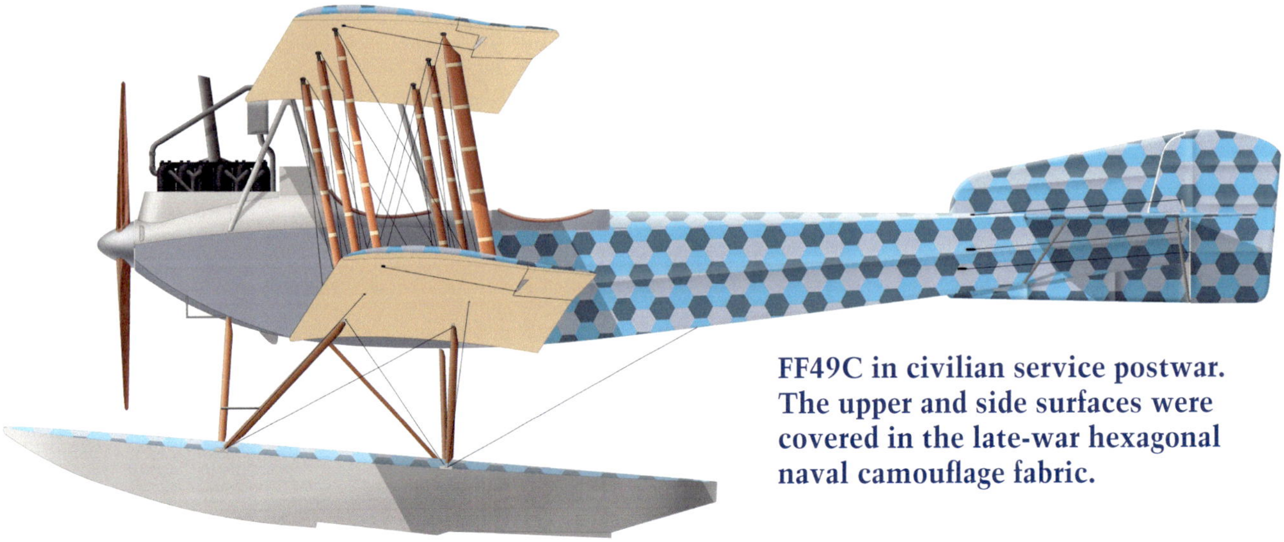

FF49C in civilian service postwar. The upper and side surfaces were covered in the late-war hexagonal naval camouflage fabric.

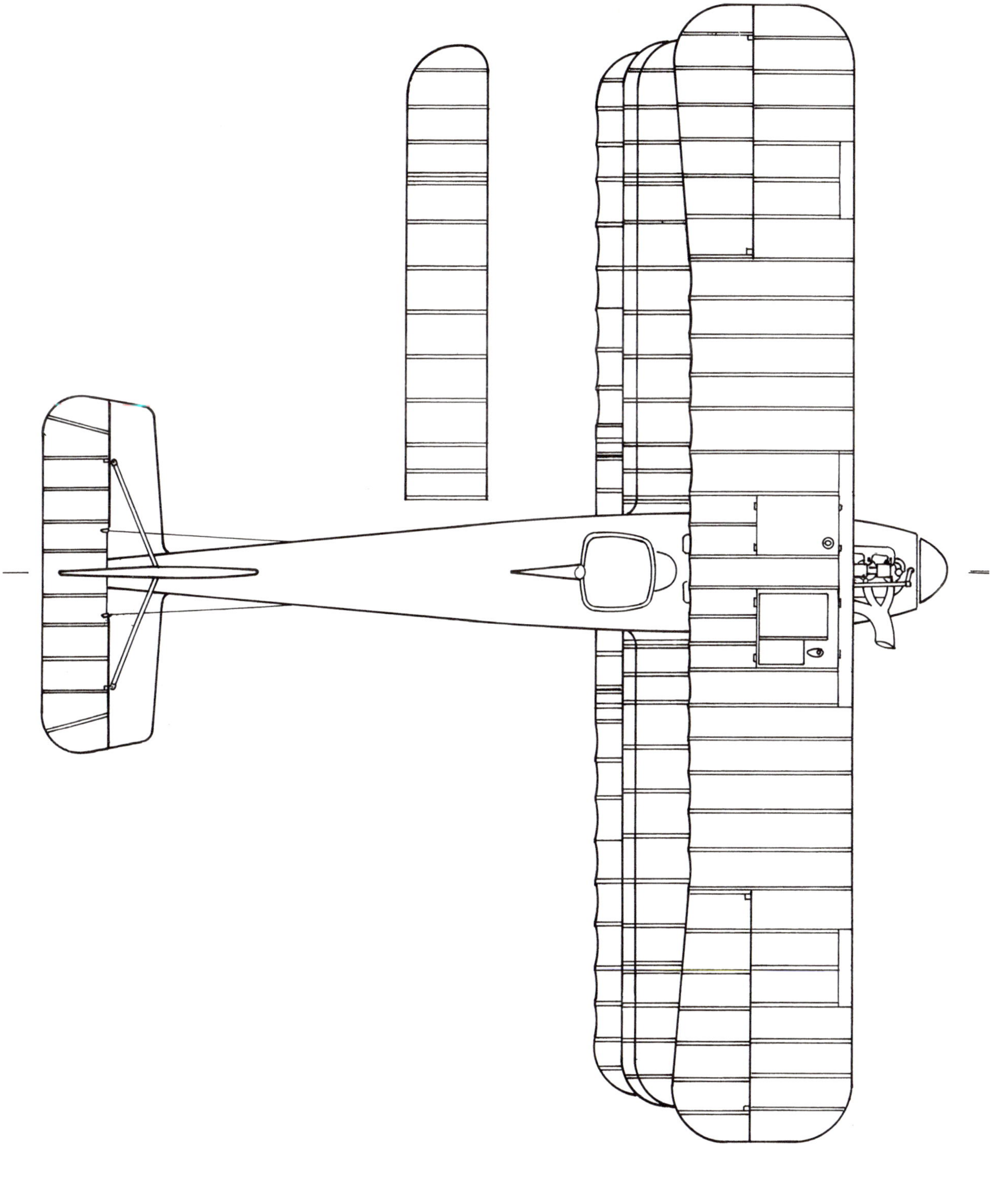

Afterword

Above: Friedrichshafen FF33E *Wölfchen* and her crew photographed for propaganda purposes on 6 March 1918 on the raider *Wolf. Ltn. zur See d.R.* M.A. Stein (left) and *Oberflugmeister der See* Paul Fabeck (right) were the crew and each could handle both positions. Fabeck had just completed his pilot's exam when he joined Wolf's crew on 9 November 1916. From July 1918 Farber served in *Flandern* II, Ostende. The wireless reel is clearly visible forward of the fuselage cross.

Unlike Fokker or Albatros, Friedrichshafen is not a famous name in WWI aviation. This is primarily due to the types of aircraft Friedrichshafen produced, the reconnaissance seaplanes and night bombers that were so important to the war effort yet so unglamorous. Despite the largely unsung nature of Friedrichshafen's aircraft, there was one Friedrichshafen aircraft that became justly famous. Appropriately, it was an FF33E, one of Friedrichshafen's most successful and widely-produced and used reconnaissance floatplanes – and generally one of the most anonymous.

This aircraft, FF33E Marine Number 841, was embarked on the German auxiliary cruiser/merchant raider *SMS Wolf* that sailed from Kiel on 24 November 1916. Appropriately, the seaplane was soon named *Wölfchen* (Wolf Cub). While aboard *Wolf*, *Wölfchen* became the only seaplane to serve in three oceans – the Atlantic, Indian, and Pacific – during the war and was a vital part of *Wolf's* astonishing success.

The *Wolf* was a modern, 5,809 gross ton, 11-knot, coal-burning cargo ship armed with seven 15 cm SK guns, four torpedo tubes, and a number of smaller guns. Equally important, *Wolf* carried 465 anti-shipping mines.

After a serious coal fire *Wolf* had to return to Kiel. The fire was extinguished and repairs hurriedly made, then *Wolf* departed again on 30 November, returning victoriously on 24 February 1918 after having sank, mined, or captured 28 Allied ships. Upon her return *Wolf* was full of valuable cargo such as raw rubber greatly needed by German industry that was not generally available to Germany due to

Above: The raider *Wolf* arriving at Kiel after successfully completing her 15-month raiding voyage. The starboard wings of *Wölfchen* are just visible amid ship and another seaplane flies overhead (above 'x').

Below: The crew of the raider *Wolf*.

Above: After *Wolf's* successful voyage, *Fregattenkapitän* Karl Nerger, her captain, became a national hero in Germany. Here he proudly wears the *Pour le Mérite*.

Above: Advertisement for the Friedrichshafen company as suppliers of land and sea aircraft to both the *Marine* and the *Armee* as well as world record holders.

Promoted Admiral, Karl-August Nerger was the head of German Naval Propaganda during WW2. In 1990, a war photographer, Karl-Heinz Buchholz, told Michael Schmeelke that Nerger gave his team a wealth of human, almost fatherly advice, although there was no preferential treatment.

Left: *Wölfchen* and her crew photographed 6 March 1918 for commemorative purposes on the raider *Wolf*. Stein and Fabeck shared piloting duties in *Wölfchen.*

the Royal Navy's distant blockade.

Much of *Wolf's* success was due to *Wölfchen*, which acted as *Wolf's* long-range eyes. *Wölfchen* enabled the raider to search far beyond the normal line of sight from the ship, a vital capability in those days before radar. *Wölfchen* carried a wireless transmitter and receiver to communicate with Wolf over long distances, which enabled *Wölfchen* to direct *Wolf* to intercept targets. *Wölfchen* could also carry small bombs to threaten merchant ships with sinking and bring them to a halt until *Wolf* could arrive to capture the ship and crew.

Right: The success of *Wolf* was a huge morale booster at a time when German civilians were starving and German industry was very short of vital raw materials. This poster advertises a "German Air War Booty Exhibition" in Munich. At the top of the poster is a stylized image of the *Wölfchen*, and at the bottom a representation of some of the native peoples supposably seeing an airplane for the first time. This was artistic license because *Wölfchen* flew her missions over water out of the sight of land. However, the composition does make a more striking image than *Wölfchen* flying over an empty sea. Although Hollywood has a bad reputation for this kind of over-dramatization, it was hardly the first to do so.

The poster text reads: "German Airwar Spoils Exhibition, Munich 1918, Hall 3 in the Arena at the Exhibition Park on the Thereshien Heights every Saturday in any weather. Great Military concert. In poor weather Streichkonzert in the Hall of Honor. In good weather Regimental Music in the tribunal area of the Arena. Price of admission 60 Pfennigs. Children and Military half price.

Above: Friedrichshafen FF33E *Wölfchen* beached. This is the aircraft used for propaganda purposes after *Wolf* returned to Germany, and differs in many details from *Wölfchen* as used during the voyage, including smaller insignia, outward-tilted exhaust, blunt-ended struts, etc. This is thought to be a different aircraft entirely but may have been a rebuilt *Wölfchen*. (Peter M. Bowers Collection, Museum of Flight)

Above: Friedrichshafen FF33E *Wölfchen* on the raider *Wolf*. At this time *Wölfchen* had been tidied up and national insignia repainted on her fuselage and wing surfaces. Most of her voyage, she flew without markings of any kind. The deck is crowded with prisoners and internees taken from the merchant ships that *Wolf* had sunk or captured. This photo gives some idea of the very limited space available for deck handling and storage of *Wölfchen*.

Above & Below: The crew perparing to put FF33E *Wölfchen* over the side to fly to Kiel on 6 March 1918. Fabeck is in the pilot's cockpit and Stein is checking the hoisting rig. *Wölfchen* is painted on the engine cowling as well as the aft fuselage.

Above: The erstaz FF33E *Wölfchen*. The aircraft's original fabric quickly rotted in the severe maritime operating environment and it had to be recovered several times during its voyage. Captured silk was the only available material that was even remotely suitable, and for much of its voyage *Wölfchen* was covered with silk and painted light deck gray.

Wölfchen was the first aicraft to go to sea based on a merchant cruiser. This created a uniquely challenging operational environment for a fragile, wood and fabric airplane. To make the situation even more difficult, *Wölfchen* often had to be dismantled and hidden to prevent *Wolf* being identified as the raider the Royal Navy was pursuing, and the FF33E was not designed for that. Nor was there an abundance of room aboard *Wolf* for that purpose nor to assemble and dismantle *Wölfchen,* a situation that worsened as *Wolf* had more successes and stored more war booty and prisoners aboard.

The crew of *Wolf*, commanded by *Fregattenkapitän* Karl Nerger, and *Wölfchen's* crew had to jointly develop not only the handling procedures to support the aircraft, but simultaneously had to develop the operational processes and tactics for this unprecedented voyage. Furthermore, the crew had to improvise rebuilding *Wölfchen* at sea from available materials after an operational accident caused by operating *Wölfchen* in heavy seas that caused the float struts to collapse on landing, the seaplane barely being rescued from sinking. A new starboard lower wing had to be built from scratch the engine rebuilt, and the entire aircraft covered with new fabric (at one time captured silk was used), all while at sea, but this was handled successfully and *Wölfchen* continued to fly.

The success of this dynamic duo led the naval aviation authorities to draw up specifications for a more compact reconnaissance floatplane that was more robust and much easier and faster to ready for flight and store afterward. In early March 1918, only a few weeks after Wolf's return, Friedrichshafen received an order to build three examples of a shipboard that could meet the Navy's requirements. The resulting design was the two-bay FF64 with wings that could quickly be folded for storage and spread for flight.

Meanwhile, the *Wolf* and *Wölfchen* and their crews were celebrated throughout Germany for the success of their fantastic voyage, *Wölfchen* thereby becoming the most famous Friedrichshafen aircraft.

Wolfchen's Voyage Flight Summary

Month	# Flights	Flt. Hrs.	Flt. Min.
February	3	2	35
March	13*	11	55
April	1	1	35
May	4	2	10
June	2	3	10
July	7	6	40
August	11	12	20
September	4	8	5
October	4	7	30
November	3	2	20
December	4	3	5
Total	**56**	**61**	**25**

Notes: All flights in 1917. *The War Diary of *Wolf* states only 11 flights in March; the other data is from *Wölfchen's* operational report.

Made in United States
Troutdale, OR
07/26/2025

33171686R00188